The Game of Science

5th Edition

Garvin McCain

The University of Texas at Arlington

Erwin M. Segal

State University of New York at Buffalo

Brooks/Cole Publishing Company

Pacific Grove, California

Brooks/Cole Publishing Company
A Division of Wadsworth, Inc.

Printed in the United States of America

10 9 8 7 6 5 4 3 2 1

Library of Congress Cataloging-in-Publication Data

McCain, Garvin, [date]
The game of science.

Bibliography: p.
Includes index.
1. Science. I. Segal, Erwin M., [date].
II. Title.
Q162.M45 1988 500 87-22396
ISBN 0-534-09072-9

Sponsoring Editor: *Philip L. Curson*
Editorial Assistant: *Amy Mayfield*
Production Editor: *Sue Ewing*
Production Assistant: *Linda Loba*
Manuscript Editor: *Andrew J. Hill*
Permissions Editor: *Carline Haga*
Interior and Cover Design: *Lisa Thompson*
Interior and Cover Cartoons: *Tony Hall*
Art Coordinator: *Sue C. Howard*
Interior Illustration: *Judith L. Macdonald*
Typesetting: *Kachina Typesetting, Inc., Tempe, Arizona*
Printing and Binding: *Malloy Lithography, Inc., Ann Arbor, Michigan*

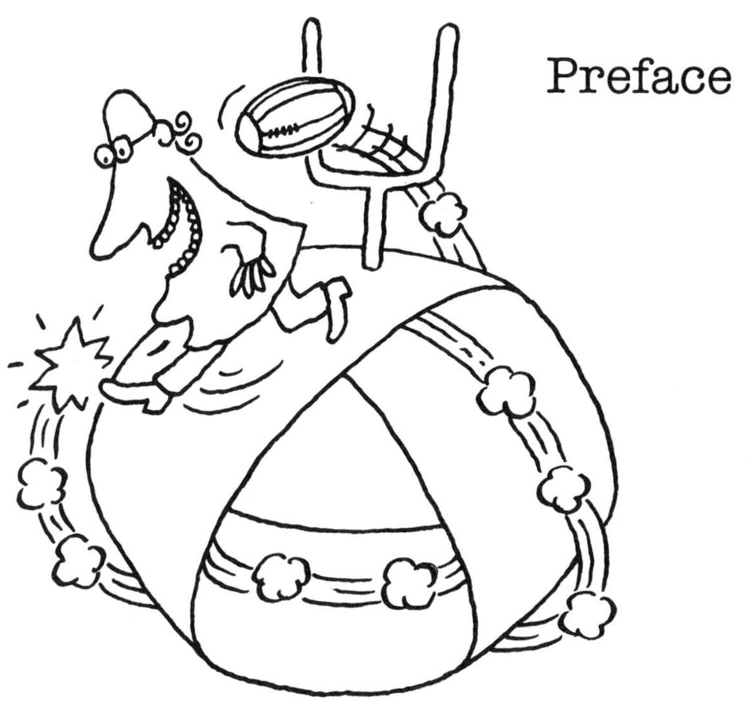

Preface

Science is a dominant theme in our culture. Since it touches almost every facet of our life, educated people need at least some acquaintance with its structure and operation. They should also have an understanding of the subculture in which scientists live and the kinds of people they are. An understanding of general characteristics of science as well as specific scientific concepts is easier to attain if one knows something about the things that excite and frustrate the scientist.

This book is written for the intelligent student or lay person whose acquaintance with science is superficial; for the person who has been presented with science as a musty storehouse of dried facts; for the person who sees the principal objective of science as the production of gadgets; and for the person who views the scientist as some sort of magician. The book can be used to supplement a course in any science, to accompany any course that attempts to give an understanding of the modern world, or—independently of any course—simply to provide a better understanding of science. We hope this book will lead readers to a broader perspective on scientific attitudes and a more realistic view of what science is, who scientists are, and what they do. It will give them an awareness and understanding of the relationship between science and our culture and an appreciation of the roles science may play in our future. In addition, readers may learn to appreciate the relationship between scientific views and some of the values and philosophies that are pervasive in our culture.

We have tried to present in this book an accurate and up-to-date picture of the scientific community and the people who populate it. That population has in recent years come to comprise more and more women, a phenomenon that we discuss at some length in Chapter 8. This increasing role of women in the scientific subculture is not an unique incident but, rather, part of the trend evident in all segments of society as more women enter traditionally male-dominated fields and make significant contributions. In discussing these changes and contributions, however, we are faced with a language that is implicitly sexist, one that uses male nouns or pronouns in referring to unspecified individuals. To offset this built-in bias, we have adopted the policy of using plural nouns and pronouns whenever possible and, when absolutely necessary, alternating *he* and *she*. This policy is far from being ideal, but it is at least an acknowledgment of the inadequacy of our language in treating half of the human race equally.

We have also tried to make the book entertaining as well as informative. Our approach is usually informal. We feel, as do many other scientists, that we shouldn't take ourselves too seriously. As the reader may observe, we see science as a delightful pastime rather than as a grim and dreary way to earn a living.

A few of our comments obviously represent essentially unconfirmed personal opinions, particularly those concerning scientists themselves. Since the purpose of the book is to represent science as some scientists see it, we believe that these journalistic expressions are proper and fitting. Furthermore, since no one can write without making value judgments, we make them blatantly and without apology, rather than attempting to deceive the reader.

Many activities and disciplines fall within the broad realm of science. To keep our subject in focus we have emphasized one particular aspect of the scientific effort—the aspect that is usually called *basic science,* the acquisition of knowledge for its own sake rather than for practical reasons. The concept of basic science as expressed in this book does not apply to formal science, including logic and mathematics. Empirical science uses rather than investigates these disciplines. Unless otherwise qualified, basic empirical science is implied whenever the word *science* is used.

We attempt to emphasize basic science in general rather than any one of the basic sciences. The game of science is played on many different fields using the same or similar rules; thus we hope the context of this book is applicable equally to physical, biological, and social sciences. We feel that knowledge of the general rules of science aids one in understanding a particular science. By using illustrations from varied fields we hope to emphasize, by example, the unity of the sciences.

Examples in the book were chosen to illustrate points rather than to provide comprehensive coverage. They have been drawn from friends and acquaintances as well as from many histories of science. We have deliberately chosen them from several different disciplines in order to show both similarities and differences within the scientific community. Although a substantial number were taken from some of the classical accomplishments of scientists, they do not represent any particular era; they simply illustrate the idea at hand.

When we present an example we generally identify the major figures connected with it in order to personalize it and to acquaint the reader with a few classical scientists. On occasion we identify the most popular, rather than the most insightful, contributor; we apologize in advance. The specific citations are not given in the text, because we don't want to divert the reader's attention. However, a number of pertinent references are described briefly at the end of the book.

A book such as this necessarily contains a number of technical terms that may be defined somewhat differently by different scientists; in most instances we have used definitions that are acceptable to a large number of scientists. We generally define these terms explicitly when they are introduced.

It would be an understatement to say that a great many things have happened to the scientific world since we wrote the first edition of this book. One of the fundamental changes has been in the attitude toward science of the public, the academic world, government, and indeed segments of the scientific community. These changes will be discussed at some length in Chapter 9. Attitudes are ordinarily reflected in some action, and attitudes toward science are no exception.

In spite of the scientific and personal changes of the past few years, the basic attitudes we expressed in the original edition have not changed greatly. This lack of change does not reflect simpleminded unawareness. Attitudes and analyses of science have been discussed by scientists for over 500 years. While both attitudes and analyses have been modified over the years, the modifications have reflected changes within science rather than from external pressures.

Readers will note that in passing we have given the back of our hand to a number of groups. Most of these unkind cuts reflect commonly held attitudes among scientists of our acquaintance. Our own attitudes are not detached from all personal acquaintance with the groups involved. We have between us a number of years' experience in science, the military, business, administration, and applied-science worlds. All of the maligned groups play an essential part of the modern scene. Their greatest fault is a narrow, self-satisfied view of themselves. Their resemblance to people, and scientists in particular, is rather striking.

Both of us owe much of the content of this book to what we have learned from teachers, colleagues, and the printed page. If this book has value, the credit should go to them. In particular we would like to thank Professors E. John Capaldi, Charles N. Cofer, Herbert Feigl, James J. Jenkins, and Paul E. Meehl for directly creating our interest in and contributing to our knowledge of the game of science. Without personal contact with such stimulating scholars we would never have begun, let alone finished, this book.

Certain scholars whom we have never met have helped make us aware of the real importance of the history of science in understanding the game. The most important of these scholars are Bertrand Russell, James Conant, and Thomas Kuhn.

Autumn Stanley helped us with the section on some of the contributions of women to the advancement and development of science, and for this we owe her our deepest gratitude.

Over the nearly 20 years since the first edition, there have been many internal and external changes. Both authors have changed their research areas, one more drastically than the other. Our knowledge of, and conclusions about, many of the topics have changed in varying degrees. The only thing that really hasn't changed is our enthusiasm for, and genuine pleasure in talking about, writing about, and participating in the game.

We would like to thank the instructors who reviewed the fifth edition for their helpful criticism and suggestions: Dr. Leo Gerulaitis, Oakland University; Dr. John Kmetz, Kean College of New Jersey; Dr. Steven Pollock, Moorpark College; and Dr. Robert Stewart, Oakland University.

We would also like to thank the reviewers of the first, second, third, and fourth editions: D. Murray Alexander, DeAnza College; Barry Anderson, Portland State College; William T. Barry, Gonzaga University; Gordon Bigelow, University of Hawaii at Manoa; Frank R. Blume, San Bernardino Valley College; James N. Bowen, University of Texas, Arlington; Herb Bryce, Seattle Central Community College; Kenneth M. Coffelt, Tarrant County Junior College; Diane Fairbank, University of Texas; Edward I. Fry, Southern Methodist University; Dr. Richard S. Goldberg, California State University-Northridge; Raymond E. Gotthold, Carlmont High School; Dr. David Graeven, California State University-Hayward; Paul T. Heyne, Southern Methodist University; James B. Hickman, West Virginia University; Edward J. Kormondy, Oberlin College; Louis I. Kuslan, Southern Connecticut State College; Peter G. Polson, University of Colorado; Louis E. Price, University of New Mexico; Brent M. Rutherford, York University; C. W. Scharf, University of Texas, Arlington; Dr. Allen B. Schlesinger, Creighton University; Persis T. Sturges, California State University, Chico; Barbara Tabachnick, California State University, Northridge; Paul E. Trejo, De Anza College; Anthony Trujillo, San Joaquin Delta College; Edward L. Walker, University of Michigan; Jules Wanderer, University of Colorado; Dr. Dale Wise, San Jose State University; and Joseph Zucca, Carlmont High School. We would also like to thank Kerry Barnes and Betty Jean McKnight, students at the University of Texas at Arlington. Brian McCain, at Schreiner Institute, and Janet L. Mistler, at the State University of New York at Buffalo, who read the book at the manuscript stage; and Gladys Wehner, Gloria Myers, and Karen Ensminger Hoyer, who somehow converted our illegible scrawls into readable type. We very much appreciate that assistance. Finally, we would like to thank Anne McCain Coffman and Carol Segal, who were incisive in their criticism and tolerant of our excesses.

We alone, of course, are responsible for any errors, misunderstandings, and faulty generalizations contained in the book. We can't blame the reviewers, because at times we were obstinate and ignored their advice. As co-authors we share equally the responsibility for the book. The order of the listing of this responsibility was determined by an ace over a ten.

Erwin M. Segal
Garvin McCain

Contents

1
An Introduction to the Game

***W**hy don't you believe in witches? That question may seem ridiculous; but our ancestors, who were probably as bright as we are, did believe in them and acted accordingly. Why are we so different and superior? The evidence for and against witches is no better today than it was 400 years ago. For us, it is almost impossible to believe in witches; for our ancestors, it was equally difficult to deny their existence. Our new beliefs are due, in part, to the development of "scientific attitudes."*

Most people don't actually have scientific attitudes, but our cultural beliefs are influenced by the fact that a small but important minority does have them. Although the scientific attitude affects beliefs, specific beliefs are not its most important product. According to Bertrand Russell, ". . . it is not *what the man of science believes that distinguishes him, but how* and *why* he believes it. His beliefs are tentative, not dogmatic; They are based on evidence, not authority." [emphasis added]

This opening was written for the first edition of this book about twenty years ago. In succeeding editions we first modified it because the beliefs of many Americans had changed; then, we dropped it altogether. We made these changes because of the shifting public view of the mysterious, the occult, and the weird. Our own views have not changed. Witchcraft, mysticism, astrology, divining, prophetic dreams, ESP, and the like have been around for a long time, some reaching back as far as recorded history. Although no one has ever been able to establish reliable evidence for mysticism, it has brought consistent monetary rewards both for mystics and nonbelievers cynical enough to capitalize on it.

The situation today is rather a strange one. Both mysticism and science are held in high regard by the public. We suspect that many individuals hold both these views at the same time. Just how this is possible is worthy of study in itself. Since 1973 (with the exception of one year) science has been rated second only to medicine in public confidence. It has far outdistanced other categories such as the Supreme Court, religion, the military, and especially the executive and legislative branches of government.

At the same time, a large portion of the adult public would be willing to restrain scientific inquiry into such areas as: enabling parents to select the sex of their child (62%), creating new life forms (65%), and even enabling most people to live past 100 (32%). Today papers such as the *National Enquirer* and *National Observer* flourish, as do films (and TV) on the Bermuda Triangle, UFOs, ghosts, and psychics. Astrology columns are carried in most daily newspapers.

One new event has been the rise of religious fundamentalism. Interestingly, this is widespread and far-reaching, occurring among many religious groups including Christianity, Islam, and Judaism. Just how these religions could all be absolutely correct we leave others to decide. They do agree in one respect: each one holds that its teachings are without error and science must agree with those teachings. Marxist "science," as we shall see, has long taken the same position, but its claims have gotten a bit frayed.

In this book we explore many aspects of science. We use the metaphor of a game because science is challenging, entertaining, and governed by rules. Most of all, science is intrinsically interesting, especially when presented well and applied properly.

A Case Study from the Game

William Harvey is an important figure in the history of science because he solved an important puzzle. In the early 17th century, he discovered that blood circulates from the arteries through the veins and back to the heart, which, in turn, pumps the blood back again. People had been interested in blood and the heart since antiquity, but the beliefs about these organs contained as much myth as fact. Different physicians and scientists believed one or more of the following propositions: The arteries work like bellows, drawing in air when they expanded and forcing out dark vapors when they contracted. Others thought the blood oscillates through the arteries and veins in the same manner as air goes up and down the windpipe. Another group believed that the body supports two blood systems: the veinal system, which distributes food to the body; and the arterial system, which distributes "vital spirits."

Harvey decided to try to figure out what was true and what was myth. His method was to gather observable facts, which he tried to assemble into a coherent story. Any "facts" that were not confirmed by observation or were incompatible with a story based upon those observations were rejected. He believed that the pieces of information ought to fit together like a puzzle. If a piece didn't fit, it wouldn't be considered part of the finished picture.

Harvey confirmed the following observation: if an artery were clamped and isolated and then cut lengthwise, it contained blood—and nothing but blood. It did not contain air or dark vapors of any sort. The same was true for a vein. When an artery was cut in a living organism the blood pulsed out, but contrary to what was believed by the second-century physician Galen, it pulsed out when the artery was dilated and not when it was contracted. Thus, the blood flow caused the expansion of the artery. The contraction of the artery was due to the movement of blood emptying the artery. Also Harvey saw that the artery, upon expansion, did not draw in air through the wound, as a broken bellows might. Therefore, the artery was a passive transporter of blood, dilated by the impulse of the blood; the artery is not the source of the blood's movement. The heart, on the other hand, contracted at the time of the blood spurt. This action pumps the blood. Harvey also saw that the direction of the blood flow seemed to be regulated by structures that functioned as valves. They allowed the blood to flow through the vessels in only one direction. Upon contraction, the heart forced blood out into the arteries; when the heart relaxed, blood flowed in from the veins. Harvey concludes:

> I finally saw that the blood, forced by the action of the left ventrical into the arteries was distributed to the body at large, and its several parts, in the same manner as it is sent through the lungs, impelled by the right ventrical with the pulmonary artery, and that it then passed through the veins and along the vena cava, and so round to the left ventricle. . . . This motion we may be allowed to call circular.

Several important, currently accepted theories derive from the work of William Harvey. The arteries and veins are parts of a single blood system. The heart functions as a pump, and it pumps the blood through the arteries into the veins. The blood vessels do not contain air or vapors. The pulse is due to the activity of the heart and is not caused by the arteries. Beliefs firmly held for many years by many people were rejected because William Harvey solved a scientific puzzle. Such puzzle solving is a major part of the game of science.

Playing and Observing the Game

Using the game of science Harvey generated conclusions that changed the course of biology and medicine. The playing of the game of science has created the modern world.

This book can be seen as a first attempt to answer the questions What is the game of science? Who plays it? Why do they play it? When, where, and how do they play it? And how does one learn to play the game of science?

Why does one play any game? The wide variety of attractions to be found in games is shown by the following list:

1. diversion (charades, solitaire)
2. amusement (catch, pin the tail on the donkey)
3. competition (football, tennis, Monopoly)
4. intellectual stimulation (chess, bridge, crossword puzzles)
5. social interaction (post office, dancing, charades)
6. completion (jigsaw puzzles, solitaire, crossword puzzles)
7. chance (roulette, poker)
8. strategy (chess, tic-tac-toe, war games)
9. personal enhancement (keeping up with the Joneses, king of the mountain, chicken, follow the leader)
10. emotional gratification (hunting, mountain climbing, sky diving, necking)
11. curiosity (spelunking, touring, bird watching)

Different aspects of the scientific effort contain many of the attractions on this list. Because of the similarities between the attractions of science and those of a diversity of games, we can consider science a game. One qualification ought to be made, however; although we view science as a game, we note that it is a game played by professionals, and, like all games played by professionals, at times it entails activities that are tedious rather than amusing.

Curiosity is probably the most important motivation scientists have. Like spelunkers exploring a cave, scientists eagerly seek new information. Their search for new information is motivated primarily by their curiosity. Because of this, they design experiments, collect data, and then pass on to new vistas.

The world of scientists is uncertain and incomplete. They are faced with a myriad of bits and pieces of data and ways of interpreting that data. One principal goal is to impose order on chaos. The major difference between putting together a jigsaw puzzle and organizing the elements of science is that the scientist's puzzle is multidimensional and is never completed. Some pieces never fit, and many pieces are always missing.

Other aspects of games are present in scientific activity. Social interactions among scientists quite often lead to the refinement of ideas. Many discoveries come to the prepared scientist by chance. Despite this, most scientific advances come from the strategic use of the scientist's weapons, and many scientists compete with one another in an attempt to get their ideas accepted by others.

For many scientists, one strong motive that is related to competition and personal enhancement is the desire for recognition. One form this ambition takes is seeking to win the Nobel Prize. It's clear, for example, that James Watson was competing with Linus Pauling to understand DNA first and win the award. This is not so different from athletes' going to the Olympic Games to win gold medals. One last motive that readers have suggested is not usually found in games—improving the world. Although the applied scientist is likely to consider this noble but distant goal, it is quite often subordinated to the everyday goals mentioned above. In fact, some scientists were actually disappointed when Jonas Salk and Albert Sabin discovered a vaccine against polio, because they had hoped for the glory of discovery for themselves.

Scientists working on a problem become very enthusiastic and emotionally involved. To understand a colleague's idea, to generate an idea, and to speculate on possible alternatives are all rich sources of intellectual stimulation. Rather than trivial, this stimulation is one of the most pervasive sources of pleasure for scientists. The game of science can be all absorbing; it can define a world into which scientists escape, body and soul. They can forget the dull, humdrum everyday routine where they, like the rest of us, spend a good deal of their time. The passion of intellectual challenge can compete successfully with any other human passion. Intellect cannot be separated from emotion. Progress toward reaching scientific goals can arouse one to a feverish pitch that may physically drain the scientist for a short time, but total commitment and involvement can often be sustained for many years.

This book describes the game of science and is organized around that theme. We have just discussed some of the reasons why the game is played. In Chapters 2 and 3 we emphasize what kind of a game science is and how both spectators and participants view the game. Chapters 4, 5, and 6 spell out the way the game is played, its rules, and the varied activities of its participating scientists. Chapter 7 describes some of the factors involved in the birth of new scientific games. Chapter 8 describes who the participants are, what motivates them, what some of their values are, what they believe, and how they become scientists. Chapters 9 and 10 discuss the results of the game—what it does for and to society.

Some readers have objected to characterizing science as a game. Any endeavor that leads to bombs powerful enough to have killed 200,000 people at one time is too serious to be called a game. We do not deny the seriousness of science—it can be an extremely serious endeavor that has widespread, lasting, and even devastating effects on people and their environment. However, one of our biases is that human activities—especially those to which people are committed—have to be justified as ends in themselves as well as means toward other ends, since a lifetime of work must be fulfilling in itself and not simply have worthwhile ends. We note the similarities of scientists' motives for doing what they do to those of individuals involved in less serious games. In fact, we believe that the game analogy applies to all worthwhile human enterprise. We just hope that those who play serious games play them well.

To play or even to observe the game of science requires some effort and a skill that takes time to perfect. As with many other games, beginning play may take one of two forms, whether one is a player or an observer. The going may be tough and seem not worth the effort, or one may play superficially and feel that the game is trivially easy and not worth learning in detail. We believe that neither need be the case with this book. One of the authors has often read an article or a book and found it to be trivial. After being called a blockhead (or worse) by a colleague (at times the other author), he has found that a rereading of the article revealed it to be very valuable. On other occasions we have read an article and found it extremely hard to comprehend. In many of these cases rereading has simplified the article immensely. We do not understand how this process works, but we profit from our experiences. We believe science is an engaging game that can be rewarding to both players and observers when they understand what is going on. We think this book has captured some of that drama. To make the reading easier, we have included the definition of many technical terms with their first occurrence in the text. These terms are all listed in the index.

No one is educated who does not understand something about the game of science. We hope to educate as well as entertain. If you have not yet done so, please read the preface so that you are aware of the framework within which we wrote the book.

We do have a profound admiration for scientific achievements, and, when we mention a conflict between science and external forces, our prejudice is typically on the side of science. This does not mean, however, that we are unaware of the human foibles and failures within science.

Science is not infallible. Until 1962, chemists were convinced (and the texts said so) that the "noble" gases (neon, argon, krypton) would not form compounds. Accepted chemical theories recognized this "fact." Unfortunately for them, a chemist who was only 31 years old at the time, Neil Bartlett, was able to produce compounds from these gases.

Those who have read Watson's *The Double Helix* or C. P. Snow's novel *The Search* (we recommend both) are aware that scientists have at least their share of human vanity, greed, lust, sloth, and indecisiveness. For another viewpoint on the double helix read Anne Sayre's *Rosalind Franklin and DNA*.

Scientists have been remarkably unsuccessful in helping nonscientists to understand science. Even captive audiences in educational systems generally depart as scientifically innocent as they were on their day of arrival.

Lest you believe that "true" science never takes a pratfall, you are invited to read the great American chemist Irving Langmuir's paper

"Pathological Science," describing the detailed work of some eminent and not-so-eminent scientists (mostly physicists) who found things that just weren't there. Yet hundreds of papers were published supporting these illusory results.

These statements are not meant to indict science or scientists but rather to recognize human limitations.

No one can write a book without making some value judgments. We affirm the view that value judgments are both proper and necessary. Clearly we made the judgment that science was worth writing about. This is no apology, merely a warning. We attempt to identify the more clearly unsupported personal views as they occur.

One final note: at the end of the book is a bibliography that includes a short description of each entry. We hope this will guide you to some interesting and pertinent reading. Also useful for term papers.

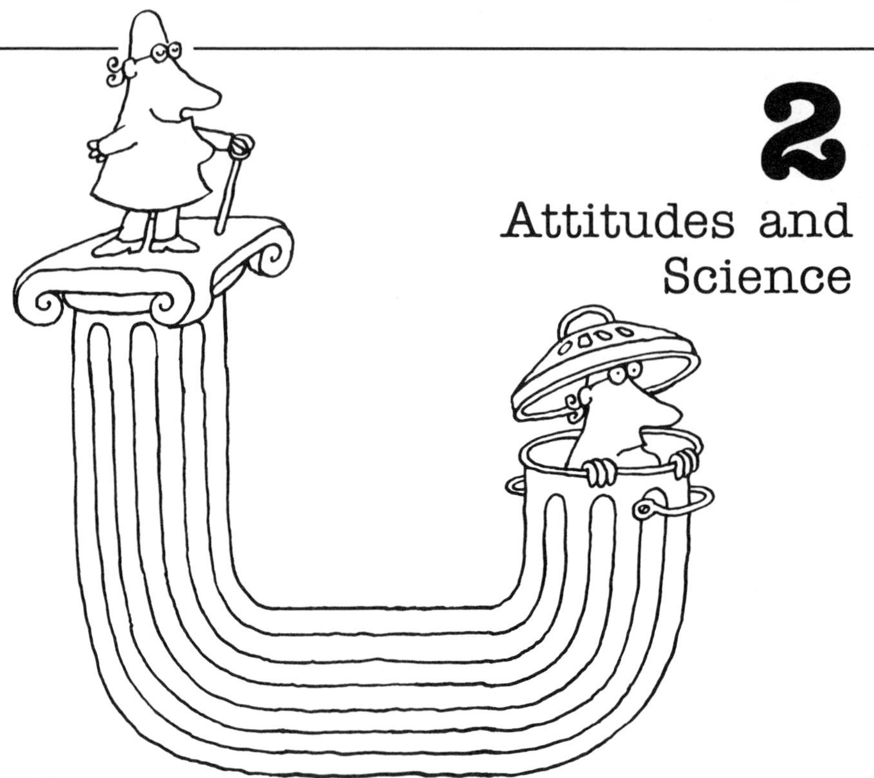

2 Attitudes and Science

Genghis Khan, Alexander the Great, Napoleon, and Hitler sowed destruction far and wide in attempts to conquer the world. Christianity, Islam, communism, and democracy have by persuasion and sometimes by force attempted to conquer the world. All have failed. The only truly universal culture is science. It would be difficult to find any person—whether scientist, skeptical of science, or ignorant of science—who has not been affected by the scientific revolution.

During the past few years we have been bombarded with so-called nonfiction about the habits of gnomes, ghosts in Amityville, carvings of an ancient astronaut in Palinque, Mexico, mysterious disappearances of boats and planes in the Bermuda Triangle, prehistoric airfields in Peru, and many other strange and mysterious happenings. These reports have been accompanied by an increase in claims by individuals that they have the supernatural ability to bend keys or forks, read minds, see things that are not in view, or mentally put pictures into film. Also during the last two decades, admittedly fictional stories about exorcists, witches, psychics, and Antichrists have been increasingly popular. Polls have shown greater and greater acceptance by the U.S. public of the truth of such stories.

Large numbers of Americans now strongly believe in possession by the devil, witches, ancient astronauts, and regular visits from creatures from outer space. Why is this so? Belief systems depend a great deal on our attitudes toward knowledge. They are also strongly influenced by cultural trends and the beliefs of our families and friends. In this chapter we will explore different attitudes toward knowledge and some of the consequences of such attitudes.

A person's particular job or profession is not the key to his or her possessing or understanding scientific attitudes. There are those trained in highly theoretical and advanced sciences, such as physics or astronomy, who not only don't know how to play the game of science but who don't even know the game is going on. On the other hand, many individuals do play the game well, even in less precise sciences such as sociology or economics, where because of the everyday familiarity of the topic, it is more difficult to maintain scientific attitudes. These attitudes are distinctly more important as a criterion of science than is the sheer amount of solid data available or the degree of development of a particular science.

Although this chapter is titled *Attitudes and Science,* not everything included in it is concerned solely with attitudes. For example, the section on "Science and the Application of Scientific Information" discusses how various people, scientists included, view the field and what they consider important, as well as how scientists focus and work on problems. These seemingly diverse topics are included in a single chapter in part because we felt that they needed to be introduced early in the book. Yet there is also a general theme. This chapter emphasizes the importance of attitude, and all of these topics do relate to attitudes or judgments rather than to specific findings.

The Importance of Attitudes

At various places and times in the human struggle, there has been a variety of dominant attitudes or modes of thought. Nationalism, theology, the exquisite glories of war, business, and racism have each at one time or another functioned as the primary yardstick for human belief systems. Consider one of these yardsticks—theology. In the 16th century theology clearly dominated the thought of Western Europe. Of course, not everyone was a theologian. For example, only a minute portion of the populace had even a foggy understanding of the ontological argument (an argument for the existence of God). But the dominant mode of thinking involved theology at some point. It is equally evident today that, although science plays a dominant role in our society, only a small minority has any serious understanding of it. On a very concrete level, almost all students accept the idea that the earth travels around the sun. But if asked why or how they know this, most people would answer "because it's true" or "I read it." We are all aware of the attempts to link science with such fascinating subjects as deodorants, gasoline additives, and false teeth. Whether or not these links are legitimate, they exemplify the advertisers' confidence that "science" has sufficient glamour and acceptance to be worth a few dollars' extra profit. Implied is the further assumption that the viewing, reading, or listening public does not understand science, so that an oblique reference to "science" will reduce them to helpless acquiescence. Whatever the effectiveness of this approach, it does reflect an assessment of popular views. Many people seem to believe that production of better monsters, tranquilizers, satellites, bunion pads, or more humane bombs is the aim of science and that these items are developed through a collection and cataloging of facts. Such "practical" items may be necessary for late, late shows, the military–industrial complex, comfort for aching feet, and the waging of wars, but their relation to science is rather remote. The popular conception of science thus emphasizes gadget production, mysterious midnight work in the laboratory, and data collection and storage. Our aim is to suggest a somewhat larger perspective regarding scientific attitudes and to present a more realistic view of what science is and what scientists do.

It needs to be emphasized that, although science affects belief, that is not its primary purpose. Science is primarily an attitude toward problem solving; as such, it often has implications far beyond the immediate data—implications that may indeed profoundly influence beliefs. For example, when Copernicus and Galileo shattered the earth's pretensions as the center of the physical universe, the impact was much greater than the astronomical consequences of accepting the sun as the center of the universe. Presumably, the Copernican concept of the universe did not change the productivity of the fields, turn wine to vinegar, or render martial and marital games less fascinating. And yet, when a long-established and firmly held belief is shattered, life can never be quite the same. At least as menacing but more formless was the thought that, if one fundamental truth had to be discarded, it might prove to be the hole in the dike through which a flood of unknown and frightening consequences might follow.

Science is very much like other human endeavors in that attitudes are of extreme importance and reach far beyond the immediate effect. Consider the doctrine of the divine right of kings. At one time this idea was widely accepted, but by the latter 1600s it was seriously challenged or rejected through most of Europe. Although the initial change of attitude involved primarily the monarch, it foreshadowed curtailment of the privileges of the aristocracy. Later consequences of this shift in attitudes included the doctrine of equality of men and the questioning of hereditary property rights. It would be absurd to assume that all of these later developments in social structure resulted solely from the overthrow of the divine-right doctrine, but the change in attitude accompanying this overthrow did play a substantial part in their development. Today it is extremely difficult to understand, much less re-create, the attitudes and implications that accompanied acceptance of the divine-right doctrine.

In summary, when we change the way we view any important aspect of our world, our attitudes toward other aspects of the world also change, because attitudes affect our perception of facts and are generally more important than specifics.

Misunderstandings about Science

There are a number of common misunderstandings about science. The following statements illustrate some of them. We'll discuss each of these misunderstandings briefly.

1. The accumulation of facts is the primary goal of science.
2. Some sciences can be described as *exact.*
3. Science is deficient because it cannot give any ultimate explanation of natural processes.
4. Science distorts reality and cannot do justice to the fullness of experience.
5. Science is concerned primarily with our practical and social needs.

1. The accumulation of facts is the primary goal of science. An important part of the game is collecting data, but mere accumulation of facts does not necessarily give birth to science. All human groups collect data, but not all humans are scientists. Consider the following vignette. As in a dream, transport yourself to a gentle isle of the South Seas. Here frolic the happy children of nature, their golden skins glowing amid glittering sands, azure seas, and lush foliage. Dominating this spectacular beauty is a mountain from which a pale trickle of smoke arises. What do the people know about this object in their midst? Quite a bit! They have amassed a wide range of data. They recognize the rumbles that precede a shower of ashes; they know that the refuse of an earlier eruption is extremely fertile; they know that ashes and lava are hot and have a distinctive odor. Legends of the village tell of fireworks bursting high above the mountain.

We could continue to catalog their knowledge indefinitely, but a simple recital of data has not led to any "scientific" understanding. The missing ingredient is a set of general principles. The inhabitants' accumulation of data is extensive, but aside from the classic line "The gods are angry tonight," they make no sense of the welter of events.

To summarize the point of this illustration, data have been accumulated in all cultures, but science is recent and flourishes in only a few. The quality and character of explanatory or organizing concepts and the relation of concepts to data are critical additional ingredients in the game of science. The concept of angry gods doesn't lead to any reliable prediction of future events; nor does it integrate specific observations into an organized pattern. In addition, angry gods won't hold still for testing.

2. Some sciences can be described as exact. There are *no* "exact" sciences! And there is no prospect that there will be exact sciences in the future! It may occur to you that mathematics is "exact." Bertrand Russell and Alfred North Whitehead, two of the most eminent mathematicians of the 20th century—possibly of all time—commented on mathematics. Russell said "Mathematics is the only science where one never knows what one is talking about nor whether what is said is true." Whitehead, in a more lyrical way, said:

> I will not go so far as to say that to construct a history of thought without profound study of the mathematical ideas of successive efforts is like omitting Hamlet from the play which is named after him; that would be claiming too much. But it is certainly analogous to cutting out the part of Ophelia. This simile is singularly exact, for Ophelia is quite essential to the play. She is very charming and a little mad.

Mathematics is interesting, but it is not science.

Science must be related to observations, and mathematics, as Russell wrote, is not derived from or dependent on observations. Mathematics is based on the logical consequences of a set of postulates. Some people call observationally based sciences "natural sciences" and mathematics and logic "formal sciences."

Statements made in natural science have only probabilistic, not exact, confirmation. Take as an example a long-established and repeatedly confirmed scientific law. Consider the distance (S) traveled by a falling body in its first two seconds of fall. The established law is presented as $S=\frac{1}{2}gt^2$ (g is a "constant" that varies with location and altitude; t is time in seconds). Actual measurement of distances (taken at St. Louis, Missouri), however, would generate a distribution of values like that in Figure 2-1. That is, repeated measures under the same conditions (as nearly identical as humanly possible) give different answers. The data conform to laws of probability that imply that one gets a particular range of results a certain percentage of the time. There is *no* way we can determine *exactly* how far a body will fall in a given amount of time under *any* specific condition. What we can do is state with a high degree of assurance that in this case 95% of the measures will fall within 0.0001 meter from the average.

In addition to their probabilistic nature, scientific statements are incomplete. Although it is generally unspoken, a qualification is implied in any scientific statement: "Based on the evidence available to me, interpreted to the best of my ability, I believe that . . ." In other words, a scientific statement is inevitably incomplete in evidence and interpretation. "Evidence available to me" leaves the scientist's statement open to revision based on new evidence. Similarly, interpretation of existing evidence may change. For example, you may be walking down the street and suddenly hear an explosive sound. You fall to the ground out of fear for your life. But, as you lie there, you may decide that it was a car backfiring. Note that the direct evidence has not changed; it has just been reinterpreted.

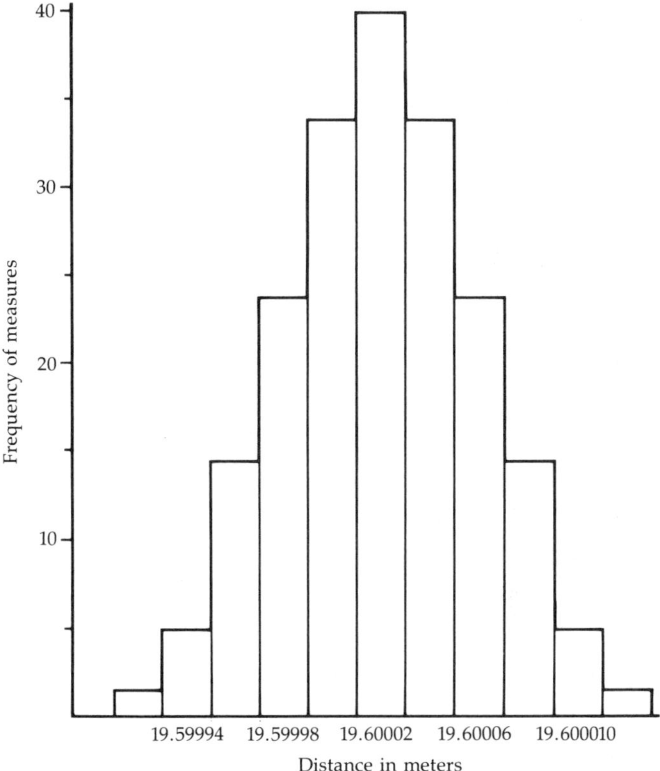

Figure 2-1. Estimated distribution of 200 measured distances traveled by a falling body in the first two seconds starting from rest.

Changes in interpretation due to reanalysis of the prior evidence, adoption of new interpretations from external sources, and new interpretations derived from added evidence are all quite common. As you saw in the introductory case study, Harvey reinterpreted some of Galen's data to make his case about blood circulation. Copernicus reinterpreted Aristotle and Ptolemy's data on the changing locations of the sun and stars, Lavoisier reinterpreted some of Priestley's data on burning objects, and Einstein reinterpreted Newton's data on planetary orbits.

Sciences, whether as well developed as physics or as recent and incomplete as sociology, have at least one characteristic in common: pronouncements are tentative. If they were not, we would have to deny the importance and even the possibility of new evidence and new thought. Such denial is contrary to the way of a scientist.

3. Science is deficient because it cannot give any ultimate explanation of natural processes. This charge is often made by people whose understanding of the methods and objectives of science can most charitably be characterized as limited. The charge is true but irrelevant. Ultimate explanations simply are not part of the objectives of science in the modern world. A quest for ultimate explanations dominated much prescientific thinking. Thousands of years of effort and thought did not lead to a resolution of any ultimate problems, such as "What is the essence of life?" or "What is the greatest good?" In contrast, the preoccupation of science with potentially solvable problems such as "Why didn't milkmaids contract smallpox?" has produced spectacular results.

As stated earlier, scientific statements are tentative. An attempt at ultimate explanations based on tentative statements could be handled only by Don Quixote or Lewis Carroll's Red Queen.

4. Science distorts reality and cannot do justice to the fullness of experience. If distorting reality means that scientists select only one small portion of phenomena for investigation at any time, then they happily plead guilty. In fact, this limitation of the area of research is a vital tactic of science. Otherwise the problems are simply unsolvable.

In any given situation, there are literally innumerable things scientists could measure or manipulate. Achievement of any reasonable answers requires them to focus on one or, at best, a few aspects of any given situation. At the same time they attempt to minimize, in some manner, uncontrolled variations in other factors that might influence the result.

The limitation of a discussion to certain aspects of the total situation occurs in any endeavor. If you are describing a baseball game, you may discuss the size of the stadium *or* the color of the uniforms *or* the eccentric motions made by the pitcher *or* the number of hits each team got *or* the team spirit exhibited *or* where the fans are from *or* the temperature at game time *or*. . . . But *no one* can discuss everything. So scientists must select the aspects of the situation they wish to discuss. They differ from nonscientists in being aware of the fact that they are selecting and systematically investigating those aspects they have selected, although they may not be aware of how arbitrary the selection is.

To aid their investigation, scientists in many instances attempt to minimize the differential influence of aspects they are not currently investigating. They do so by manipulating the environment and performing experiments. Criticism that science distorts reality often focuses on the laboratory experiment. Experiments are alleged to distort "reality," since the situation is artificial. Of course, scientists do not have as a goal the distortion of reality but, rather, the discovery of how the real world works by temporarily controlling events related to those they are currently interested in measuring.

The importance of such controls can be illustrated in the field of genetics. Genetics remained a matter of rather vague statements about "blood" and "like begets like" until Gregor Mendel and those who followed him examined the situation in carefully controlled and limited conditions. Mendel succeeded where earlier hybridizers had failed largely because he studied inheritance of single contrasting characters, carefully controlled other factors, and recorded the results in detail. This is historical evidence that certain sciences advanced more quickly to create and observe artificial situations in which only certain factors can have an influence. Nonexperimental sciences include careful detailed observation in a variety of conditions, checking for regularities in order to achieve similar insights. As you may have suspected by this time, science becomes sophisticated when it is able to create and observe artificial situations in which only certain factors can have an influence. The success of science depends on the ability to screen out the booming, buzzing confusion that confounds ordinary observation.

In addition to being charged with distorting reality, scientists are also accused of failing to do justice to the fullness of experience. This accusation is most often applied to psychology, and it usually takes the general form "People are too complicated and individually different to be studied through artificial laboratory situations." This argument is nonsense, of course. Scientists do in fact study humans from the standpoint of both underlying biological processes and observable behavior. From these standpoints they create theories attempting to account for all aspects of human character. Their knowledge is both extensive and incomplete, as might be expected.

Each person is unique. So what? Every snowflake is reputedly different, yet scientists can still make meaningful statements about moisture, size, temperature, or crystalline structure. It is also said that every single atom is different. This statement may or may not be true. It is irrelevant! We do study atomic structure, but science has little or nothing to say about the completely unique. Its primary goals are to understand how observed facts relate to one another, to formulate models that approximate the relations of these observations, and to extend predictions to new observations. Few would object if you chose to spend a career studying a single snowflake; however, the scientific gain would not seem to justify this devotion. Not that the expression of uniqueness and esthetic qualities is not worthwhile—it is just not a basic part of the scientific endeavor.

5. Science is concerned primarily with our practical and social needs. The last of our misunderstandings is the assertion that science develops in response to practical and social needs. It would be fruitless to deny that such needs play a part in science, particularly applied science. But they play only a small part; they are not the game. Unfortunately, these practical results of science are the most visible, so they are often identified as the heart of science. We know that the desire for these by-products is responsible for some scientific work and that the financial backing of other work is motivated by hopes of practical gain. For most scientists, however, science is a game played for understanding, not for practical solutions to existing problems. There is no really good evidence that simple need produces scientific advancement. If the scientific groundwork has been laid, necessity may encourage solution. For those scientists who play the game for understanding rather than for practical advantage, it is a game whose chief delight is the addition of one neatly contrived stroke that helps give form to a picture; a game affording a glimpse of what no one has conceived before; a game from which may come the ecstasy of bringing order out of chaos.

A simple glance at a few of the more important figures in science reveals the absence of practical goals. Galileo, Newton, and Einstein would certainly make any all-world science team. What practical need could have driven Galileo to astronomy, where he was criticized, rebuked, and threatened, both by academics and by the church hierarchy? Where was the practical payoff? Could Newton or Einstein have been motivated by the hopes of future space travel or weapons that would fry dull, uncomprehending masses as crisp as bacon? Utterly absurd. Nor does personal gain through either fame or money seem to have been their principal motivation. Both Newton and Einstein achieved fame at an early age but continued to work until their deaths. There is a story, which may be true, that Einstein, when invited to the Institute for Advanced Studies at Princeton, asked whether $3000 a year would be too large a salary to expect. Of course there *are* those in science whose aim is a better mousetrap or personal gain. They may achieve visibility, but rarely do they make serious contributions to the basic concepts in science.

The Nonviolent Nature of Science

Although science is often revolutionary, and the *products* of science may well terminate life on this planet, the *problems* of science are not soluble by violent means. The nonviolent character is not due to any angelic aspects or overwhelming virtues of scientists. Even if a violent outbreak were to occur and one scientist to physically beat upon another, the scientific problem would not be resolved. The simple fact is that scientific problems do not readily lend themselves to violent solutions. Nationalists can make their point and achieve territorial aspirations with a bayonet. An economic philosophy can gain a place in the sun following a violent revolution, as Lenin and Mao have demonstrated. "Conversion by the sword" has had lasting effects in many parts of the world. But scientific struggles have been relatively bloodless. Galileo and his small group of followers eventually won their struggle against the church, the strongest power of their time. There is no record that they employed violence in the contest. Today, about 350 years after the church condemned Galileo, a move is under way to reopen his trial. It's a comforting thought.

Stalin used violence, threats, and bribes to impose political standards on genetic theories in supporting Lysenkoism, a theory holding that acquired traits are hereditarily transmitted. His success was fleeting and disastrous. Why? The attempted application of unconfirmed theory brought Soviet genetic research to a virtual standstill and created difficulties in agricultural development. Scientific theories must at some point be anchored in observations. Try as you may, you can't force observable data to change; they are stubborn and eternally patient. If observers go to the stake or to Siberia, there will be others to observe; observers may remain silent, but the data become like heavily starched underwear—concealed from others but difficult to ignore.

Scientific controversies are in part settled by appeals to evidence. This evidence includes observations at some point. There are no time limits on the accumulation of evidence, since controversies can be reopened or continued indefinitely. Typically, settlement of a controversy takes place over a period of time; personal prides and prejudices are slowly dissipated and are not likely to lead to violence. Elsewhere we mention that resolution of the stellar parallax problem took longer than a root canal—295 years to be exact.

Controversies are generally resolved when scientists submit their findings and speculations to a jury of their peers. Although these peers make mistakes, they allow continuous and unlimited appeals of adverse judgments. Like uncertain shoppers, scientists are rarely forced to make an immediate conclusion, and when a transaction has been completed, the "merchandise" may still be returned. Such a system can be magnificently frustrating; yet controversies are ordinarily resolved without violence.

Hostility toward Science

Despite its nonviolent nature, its glamour, and its importance—or perhaps because of these qualities—science has engendered underlying streaks of hostility in practically all levels of our society: fundamentalist religious groups claim that science intrudes on their province and would destroy all religion; secular humanists charge that science dehumanizes people and reduces them to automatons with no special quality or purpose; conservationists accuse science of turning our environment into a shambles with no forest to visit, no animals to see, no challenges to face, no water to drink, no air to breathe; members of the New Left say science threatens their lives with the bomb, deliberately destroys cultures to save them, denies people their freedom, and doesn't allow normal human existence to proceed; politicians sneer that science never accomplishes anything and is a great waste of money.

Most of the hostility toward science can probably be divided into two types. First, people oppose new ideas that challenge their beliefs. Second, some people oppose the application of the conclusions and technology of scientists to some real-life situations. To make themselves absolutely consistent such individuals should forgo anesthetics, penicillin, the polio vaccine, sanitary food, and electric lights and welcome back typhus, smallpox, the death of half the infants born, and superstition. Millions upon millions of the present world population would not survive in such a world. But this is mere detail. There is, in addition, hostility toward science from the scientific community. In many cases the scientific establishment inhibits and attacks both new ideas and the scientists who present them. The reciprocal of this hostility is also present. Many scientists are hostile toward what they deem to be the rigidity and self-serving nature of much of the scientific community with its implied absolute truth and morality. These scientists are also hostile toward many real-life decisions made in the name of science. In other words, there is hostility among scientists as well as between the rest of society and scientists. Thus, the two types of hostility toward science operate in microcosm within science.

The first type of hostility has had a long and sometimes bloody history—with scientists providing the blood. However, many of today's scientists have become free to present their opinions. Physics, in particular, has become relatively safe, primarily because it is now so complex and abstract that only a very foolish lay person can even pretend comprehension. Besides, there has been so much technological confirmation of certain theories of physics in our everyday life: we have electricity, automobiles, telephones, and refrigerators. We can read about and see such delights as spaceships to the moon and hydrogen bombs. Physical scientists are given great rein, and their most vacuous pronouncements, if limited to their field, are accepted unchallenged by the lay public. Who would challenge a science that has made it possible for J. R. Ewing to appear in our living room every week?

At the other end of the spectrum, however, are sciences that are often not recognized as sufficiently abstract to deter the lay person. In addition, these sciences tend to be closely related to some vested interest. At present, sciences such as sociology, psychology, and economics pose the same sort of threat to parts of our society that Darwin's biology and Galileo's physics and astronomy did earlier. Many nonscientists "know" the answers to problems studied by social scientists through their own experiments, observations, introspections, insights, and discussions. We all have belief systems that require certain statements to be true, and we won't accept denial of these truths from anyone. One's way of life—whether Eskimo or Texan—is usually endowed with unique virtues. Even views that seem impossible to test can be held with a vigor out of all proportion to their importance. How many violent discussions originate with a comparison of sports teams, films, or actors? Such discussions generally contain a minimum of evidence and a maximum of feeling. The point is that we become emotionally involved with our beliefs, and woe to those who challenge them, no matter what the argument or evidence. Thus the social sciences, which often challenge our personal beliefs, are applauded when they agree with us and scorned when they disagree.

The scientist's role is further complicated because scientists also have their own personal nonscientific belief system. Although these belief systems may legitimately influence research directions, ideally they should not influence scientific conclusions, but the ideal is not always achieved in practice.

Whether currently articulated specific principles from the social sciences prove useful in the long run remains to be seen. However, within each of these disciplines, at least some people understand and play the game of science. In the future, many of their revolutionary hypotheses will be blandly accepted with the same lack of emotion that now accompanies a statement that the earth circles the sun.

Our modern society is somewhat different from the societies of Darwin and Galileo. Today, even in controversial areas such as the social sciences, one can speak freely. Of course, free speech may still carry fringe benefits such as denunciation in the press, sudden change of job locale, or censure by public or private bodies. For example, B. F. Skinner, an outstanding psychologist, concluded in *Beyond Freedom and Dignity* that the majority of human behavior is learned. He also proposed that, since important ethical decisions are learned on some arbitrary basis, we should deliberately shape the child's attitudes toward ends that are beneficial to society. At least one congressman attacked Skinner's book and denounced the use of federal funds for research that led to such conclusions. In a somewhat similar situation, the U.S. Senate passed a resolution condemning the first Report of the Commission on Obscenity and Pornography (the report was based on findings by a large number of reputable scientists and scholars) without reading it—in fact, even before the report was printed! In 1986, a new commission, headed by Attorney General Edwin Meese planned its conclusion before evaluating any data.

During the last few years, religious fundamentalists have been gaining influence in the school systems. One issue that has concerned them very much is the teaching of evolution in the schools. Sixty years after the Scopes "monkey trial," several states, including Texas and Louisiana, were still passing laws limiting the teaching of evolution. Several states passed laws stating that evolution could not be taught unless it was linked to a theory of creationism or creation-science. Many teachers and textbooks, to avoid controversy, eliminated any presentation whatsoever on the origin of species. This is not good educational practice. However, all is not bleak.

On June 19, 1987, the supporters of a science curriculum, which is independent of political and religious interference, had a major victory. The Supreme Court, by a 7 to 2 margin, declared the Louisiana law invalid. Associate Justice William Brennan writing for the majority stated, "Because the primary purpose of the Creationism Act is to advance a particular religious belief, the act endorses religion in violation of the First Amendment." (The First Amendment states, "Congress shall make no law respecting an establishment of religion.") Brennan said, "The act violates the establishment clause of the First Amendment because it seeks to employ the symbolic and financial support of government to achieve a religious purpose." We agree.

Texas, previous to the Supreme Court decision, had reversed its own law, and in 1986, Steven Shafersman, president of the Texas Council for Science Education, rejected biology and geology textbooks that avoided a discussion of the process of evolution. (Texas has a unique system for choosing texts. The members of the state board select a limited number of texts that may be used. Individual school boards may choose among these. Because the educators in Texas, California, and New York purchase the largest number of textbooks, their selections carry great weight with the publishers.)

Although there are these rays of sunshine, there are still many who are ready to attack an idea or its exponents without waiting to be confused by evidence or logic.

The second major source of hostility toward science is much more modern. This attack claims that science is misused in one endeavor or another. Here we have a direct conflict between ideology and the development or application of scientific knowledge. Many people in our society simply object to many of the domains in which scientific knowledge (or supposedly scientific knowledge) is being applied. Many students of science seriously object to scientific studies being sponsored by the U.S. Department of Defense. They envision the purpose of the department as the waging of war and feel that it is using scientific knowledge to enable us to kill one another more efficiently. Although some of the department's research money is not used in any direct way to prepare for or wage war, and although some of it is for basic research, to many it symbolizes destruction rather than construction; so they oppose both the department and anything that it supports. No one can deny that much of its research money is being used to develop better weapons, better delivery systems, better evaluation of military personnel, and better spying systems. In addition, there are plans to spend many billions of dollars on a Strategic Defense Initiative or "Star Wars" defense system. All of this research tends to be classified and therefore unavailable to the scientific community. Scientific research is supposed to be done in an atmosphere of free and open communication; thus there is no clear, simple relation between the Department of Defense and publicly available scientific research. However, the department is a high-technology institution. To carry out its mission, which may be seen as preserving the peace as well as waging war, it must have a solid base in the varied sciences.

Substantial funds have also been going to very expensive investigations by the National Aeronautics and Space Administration. The major stated justification for these expenses had been scientific exploration, although recently the military role of NASA has become quite explicit.

Many people have noticed the problems in our society that are not investigated for lack of funds and hold science as well as the industrial and governmental establishment responsible. There is much discussion about national priorities; a large part of the scientific community has tacitly supported the priority system that we follow.

In addition to theoretical and practical disagreements with scientists, another cause of hostility is the perception of scientists as possessing certain personal characteristics. Scientists tend to qualify their answers; for someone accustomed to a yes or no answer, this tendency can be annoying. (We'll examine the reasons for these qualified answers later.) An amusing illustration of frustration with these qualifications took place during some congressional hearings. Former Senator Edmund Muskie observed that scientists insisted on saying "On the one hand the evidence is so, but on the other hand . . ." The senator half-jokingly called for one-armed scientists. Many scientists refuse even to discuss their findings with representatives of the popular media. The public ought to be informed, but unfortunately a scientist's statements and the final news-media presentations of them are often about as closely related as fifth cousins twice removed. Both authors have had occasion to hide from their colleagues after misguided but enthusiastic media reports. The blame does not rest solely on the media. If scientists simplify, they aid the distortion and appear simpleminded to their peers. If they explain concepts and data in detail, they put the reporter through a Ph.D. program during the interview. The problem is very real. If this book is successful in giving the reader some insight into this problem, it will have achieved one of its purposes.

Scientists are often portrayed as intellectually arrogant and unwilling to listen to or consider the views of others. There is more than a germ of truth in this accusation. However, they tend to be arrogant primarily in their own field. One example of this arrogance that gained a great deal of notoriety concerned Immanuel Velikovsky, a psychiatrist who wrote the book *Worlds in Collision*. He argued that Venus is a recent addition to our solar system and that many myths can be explained by the arrival of Venus and the movement of Mars. Many scientists denounced this thesis as foolishness without reading it. After great controversy, however, some scientists did read it but still did not accept it. They then rejected it by arguing that Velikovsky's claims are inconsistent with what is known or are too vague to be useful and therefore can be presumed to be false.

Arrogance is not confined to scientists. It exists in most if not all highly professional groups. For example, the major-league baseball player may refer to less accomplished players as "bush leaguers" or "sandlotters"; these epithets reflect neither overwhelming humility nor a high regard for the less gifted. Discuss your favorite home remedy with an M.D., your suggestions for running a business with an executive, your ideas of art with a professional artist, philosophy with a philosopher. Then judge whether scientists are alone in their arrogance.

Scientists are also alleged to be irreligious, unconventional, and of doubtful loyalty. The charge of irreligion could be more accurately phrased. If the charge is changed to a lack of identification with religion, evidence shows that the description is accurate for a great portion of the active scientific community. Many scientists find it difficult to reconcile scientific attitudes and religious feelings and beliefs. The result is that a great portion of the active scientific community simply ignores organized religion.

At one time many scientists were actively antireligious because of the open conflict between religion and science. Today, religious groups no longer have the power to suppress scientific advancement. If they reject certain scientific theories or discoveries, their influence is confined largely to the sympathetic audiences of their own members, and scientists can usually ignore them. Of course not all scientists are uninterested in religion. Repeatedly, studies indicate that, although scientists active in research tend to have little or no interest in organized religion, inactive scientists tend to have at least a moderate interest in the subject.

The charge of unconventionality of scientists depends very much on one's definition of convention. Within their own subculture, scientists tend to be rather conventional—although what is accepted as conventional in the scientific subculture still allows its members substantial latitude. The "unworldly" attitude of scientists toward wealth is an example. As contrasted with the prestige of those in the business world, among their peers scientists' prestige in the scientific community bears little or no relationship to their wealth. In this respect, scientists are no different from business executives; both conform to the values of their own group.

However, conforming to the scientific subculture is not the whole story. In addition to having different values, the scientific community has a substantial number of members who may not fit any identifiable set of conventions. The demands of the scientific game are such that high levels of independence and intelligence are required of many participants. Even in these cases, however, the individuals may be not so much nonconformist as simply unconcerned with a particular set of values.

The charge regarding loyalty is difficult to answer simply. Science is not limited by international boundaries or institutional alliances. From the very early days men and women of science have maintained contact through correspondence, conventions, and laboratory visits. A scientific triumph is part of the common enterprise, without regard to its place of birth. Science is and must be a public endeavor. Scientists' whole training and way of operating cry out against secrecy. Their profession requires information in order to progress. They tend to be loyal to their profession and their colleagues. One way of looking at the problem of loyalty is to realize that all individuals have multiple and often conflicting loyalties. A father and mother's loyalty to their children, a religious believer's loyalty to a particular god, or a scientist's loyalty to the search for understanding pose different and difficult conflicts with other loyalties. It is not implied that scientists have handled themselves badly in this field. Rather, the wonder is that with all their possible conflicting loyalties they have handled themselves so well.

In short, scientists are neither as appealing as teddy bears nor as offensive as skunks. For better or worse they are products of their subculture and the types of personalities it attracts. It is interesting to note that the relation of scientists to their society is not limited by national boundaries; Soviet scientists seem to have the same abrasive qualities within their society that our local products have in ours.

In recent years a number of Soviet scientists have been harassed, barred from laboratories or jobs, and even jailed for their work or opinions outside science or for being interested in Israel. U.S. and European scientists have protested, with only limited success. If we were cold-blooded warriors we should encourage Soviet persecution of scientists. As someone has said, Hitler lost the war because he couldn't tolerate a few Jewish scientists.

Relevance

Many people see much, if not all, of science as irrelevant. Although this might not generate hostility, it does turn people off. This is an important problem, because the future of science and the development of our culture depend to a great extent on the young men and women who become scientists. Also, in a culture that is to a great extent dependent on science, all citizens should have some appreciation of science, which is impossible to attain with a turned-off attitude.

Relevance has two aspects, practical and emotional. Many people believe that scientists should work directly on practical problems. They feel that basic research, which has to do primarily with understanding and theoretical development rather than with solving immediate problems, is thus irrelevant. This is not the most important definition of irrelevance, but it should be briefly considered.

Research on the sex life of insects has been used to aid in the elimination of pests; trying to understand lightning led to the harnessing of electrical energy for all of its myriad uses; investigating filterable life forms led to the polio vaccines; thinking about what the world would look like if one were to ride on electromagnetic waves led in part to nuclear energy; having animals press bars in boxes led to helping the mentally retarded. It isn't that the applications of science are all accidental but that topics that seem to be eons apart may be related to each other. At times one finds the solution to one problem by working on another that is related to it. In addition, if scientists have a practical problem to solve, they may find that it cannot be solved directly, but that they have to solve basic problems first. The problems associated with immune mechanisms and rejection of foreign bodies have to be solved before organ transplants can be generally successful. In fact, much of the research in social sciences—which seems very theoretical and irrelevant to real issues—started with the attempt to solve practical problems.

It is not the lack of practical applications that usually turns people off from science, however. The second aspect of relevance is more an emotional problem. Some things seem dull, uninteresting, and intuitively wrong—irrelevant to what one thinks and believes. This is the irrelevance of science that really matters.

Although we are not sure why people get turned off, some reasons have been proposed. C. P. Snow has argued that two alternative cultures compete in the intellectual world—the scientific and the literary. There is minimal communication between these cultures; so, if students begin exploring one, they tend to reject the other. One of the major differences between these cultures is that the members of the scientific culture tend to be much more optimistic about what can be done for the fate of humankind, whereas many of those in the literary culture feel that nothing can be done to change our fate and that we must therefore live each existential moment for itself.

Another reason that many people tend to be turned off from science is that they may read so-called relevant literature. For example, there has been a vast amount of publicity on marijuana use. Millions of Americans have smoked pot. Both those who have and those who haven't smoked may be interested in the subject, and a lot of scientific research has been done. What do these people interested in a relevant topic find out? They may find that naive users of marijuana have less coordination of small muscles, dryness of the mouth, and a slight increase of heartbeat rate. About 90% of marijuana research is biological and biochemical in nature, and there is no description of the states of consciousness that one feels or the "high" that one achieves. Readers infer that their psychological reactions are irrelevant to science and vice versa. However, we believe that scientific methods can be successfully applied to experiential domains, although very little has been done so far.

Finally, regardless of its subject matter, scientific prose can turn anyone off. Most of it is presented as totally impersonal and unemotional. The dynamic, emotional, competitive, exciting reality of scientific discovery is presented in the passive voice, so we read that experiments are run and theories are explained, but the scientist is never identified as a participant. The problem of communication from scientists to lay persons is similar to that of any expert to those outside the field. The dreary but critical details of procedures, the complexity of interpretation, the difficulty of terminology and many other factors make the attempt to communicate on a professional level in lay terminology a nightmare.

The mass-media attempt to communicate science news has been successful to some extent. PBS has several science programs, such as *NOVA, National Geographic,* and *Wild America,* which communicate some of the flavor of certain aspects of scientific discoveries. The *New York Times* and several other newspapers have science editors and regularly publish scientific articles. *Discovery, Natural History,* and *Psychology Today* try, with some success, to present aspects of modern science to the public. One problem is that very few people see these shows or read these items. On the other hand, articles in the *Star,* the *National Observer, National Enquirer,* and the like have fantastic and distorted articles and a very wide readership.

There are inevitably points on which co-authors do not agree. The problem of communicating scientific work is one such point. We agree that clarity, simplicity, and readability are goals fervently to be sought. We disagree about the extent to which it is possible to use language understandable to nonscientists in such communication. One of us feels that much more could be clarified if a more readable style were used; the other stresses that there are limits to what a lay person can understand, no matter how clear the language.

Bertrand Russell preferred plain English and explained how he got away with using it:

> I am allowed to use plain English because everybody knows that I could use mathematical logic if I chose. Take the statement: "Some people marry their deceased wives' sisters." I can express this in language which only becomes intelligible after years of study; and this gives me freedom. I suggest to young professors that their first work should be written in a jargon only to be understood by the erudite few. With that behind them, they can ever after say what they have to say in a language "understanded of the people."

On the other hand, Russell and Whitehead did write *Principia Mathematica,* which revolutionized mathematics, although it has been said that very few people in the world—perhaps six?—have read the entire three volumes. Others, such as Stephen Jay Gould and Carl Sagan, have tried to present science to the layperson with mixed but often delightful results.

One possible solution is separate literatures for the expert and the lay reader. The principal objective would be to give the general public an understanding of what James Conant calls the "tactics and strategy of sciences." Summaries of recent findings in science and accounts of the actual processes and human beings involved should help.

The question remains: is it possible for scientists, mathematicians, engineers, or other such specialists to write in such a way as to be bright, sparkling, and understandable to people at large and at the same time convey adequate information to other specialists?

The Assumption of Simplemindedness or Omniscience of Scientists

The title of this section may seem somewhat strange, but it does represent frequently encountered attitudes. On the one hand, people often discount or scoff at scientists when a theory violates their "common sense" or threatens some cherished belief. On the other hand, the atom bomb, polio vaccine, and rocketry have surrounded scientists with an aura like that around ancient magicians. The irony of this situation is that practical products are viewed with awe, while theoretical structures may be considered hopelessly moonstruck. Interestingly enough, scientists tend to evaluate these accomplishments in the opposite order: they value theoretical contributions far above the practical products of those theories.

A good example of the lay attitude that scientists are simpleminded was provided by the man who announced in a confident bray that he could prove Einstein was wrong when he said the universe is limited by simply asking one question: "What is on the other side of it?" The simple question designed to put a scientist in his place is not exactly a new or surprising idea. The great poet Goethe attempted to refute Newton's concept of the composition of white light by looking through a prism at a white object. He could see by simple observation that white light was not composed of many different colors—it was just plain white.

In some ways, scientists can sympathetically understand and even agree with "common-sense" notions: the earth is indeed flat, and the sun does go around it; mash a toe and the pain is immediate; a chair is a solid object, not mostly empty space (easily confirmed when it meets a shin in the dark). For most everyday purposes these answers represent perfectly reasonable ways to view the world, and should you behave as if these statements were correct you would be unlikely to come to serious harm.

The scientist's way of looking at the world is also reasonable and useful. An important reason for the assumption that the scientist is "simple" is that a scientist often deals with concepts and data that go beyond immediate sensation. Remember, the scientist has a large store of observations and concepts from many sources, and they need to be integrated in some way. For example, the scientist knows that the speed of nerve impulses varies from about 0.7 meter to about 120 meters per second, depending on the type of nerve fiber. Although pain seems instantaneous, substantial evidence indicates that it is not. This evidence has been gathered by scientists using an electrical stimulus, a microsecond timer, and electronic recorders. This work is presented in most good physiology texts. The chair example mentioned above illustrates the problem of integrating observations and concepts. With the proper instruments we can observe that the wood in the chair has a definite cellular structure that is not at all apparent to the immediate senses. Can we abandon our notions of atoms and molecules derived from, and important in, other contexts just to render our chair as solid in all ways as it feels to the shin?

The object of these examples is to emphasize that the scientist playing the scientific game has different problems from those of the lay person. These problems require what may seem rather strange answers. Scientists are not unaware of the raw appearance of things. Rather, they have to make their answers consistent with a wide range of data and concepts as well as raw appearance. They may be confused or wrong concerning a particular concept, but the reasons are not likely to be apparent to the untrained eye.

"Smog, cancer, intergalactic travel, 'Star Wars' defense system, juvenile delinquency, birth defects, schizophrenia, universal education—science can solve any problem if scientists will only work on it." This is a popular version of the view of science as omniscient. Scientists themselves are generally much more cautious; they doubt their abilities more, and they place great emphasis on the difficulties of even partial solutions of these practical problems. In many instances scientists don't even know how best to try to solve a practical problem. Unfortunately there are many problems for which science has no answers at present and others for which it may never provide answers. Another aspect of solving practical problems is that the technological aspects are not enough. Even if a problem is solvable, it is far from certain that the solution will be put to use where it would do the most good. Also, the solution of a technological problem may, in itself, create new problems. This cycle of problems causes some misgivings among scientists. Most, however, are optimistic and conclude that we almost always come out way ahead.

President Harry Truman once made an observation that might well illustrate a reasonable expectation of scientists: carved on the tombstone of a Midwesterner who died in the 1800s, said Truman, were the words "He done his damndest." Our problem is that sometimes the scientist's damndest isn't enough when it comes to solving important problems.

Science and the Application of Scientific Information

The work of many individuals rests on a scientific base. Wide varieties of roles, training, and attitudes are represented. It seems useful to distinguish three groups: (1) those directly involved in what we speak of as basic science; (2) those who organize material based on the principles of science for application; and (3) those who apply the organized material to immediate problems. Very few individuals work exclusively either in basic or applied sciences; at times untangling the particular roles might not be worth the effort. In spite of overlapping, however, there are sufficient distinguishing criteria to make this classification useful. So for present purposes we will assume that there are at least three reasonably distinct roles: the scientist, the applied scientist or engineer, and the practitioner. We can identify and classify these three science-related groups best by establishing a picture of their functions and relations.

1. Basic scientists. Basic scientists are engaged in the task of articulating, deriving, and generating principles that they hope will have general explanatory power. Their particular field of research may be rather narrow, and their work may consist of attempting to fill in more specific detail within an already existing *conceptual schema*. That is, their work may depend on a set of interconnected ideas and propositions that provides a framework for consideration of theories. The conceptual schema contains both the explicit and implicit ideas that govern a scientist's thinking about a set of natural phenomena. For example, the conceptual schema may be those ideas associated with the assumption that complex organisms evolved from simpler ones. "Mopping up"—investigating a part of nature in depth and detail within a conceptual schema—absorbs more scientific effort than any other single activity. Although this activity is not the sort of work that ordinarily results in a breakthrough, it does add clarity and precision to existing principles. For example, a great portion of the work of geologists over many years has been locating, plotting, identifying, and dating various formations. Techniques have improved vastly, but the greatest result has been a rather detailed determination of the structure of the earth's outer crust. The basic principles guiding this work were developed largely in the 19th century by William Smith and Sir Charles Lyell. Only when the idea of continental drift or plate tectonics was accepted, did geological rules, efforts, and subject matter go in new directions.

Basic scientists also engage in the sport of hypothesis testing. They make a prediction from some set of assumptions and then test that prediction. For example, scientists have known for years that Cambrian formations (rock formations assumed to be 500 to 600 million years old) contained rather highly developed fossil forms. From our knowledge of evolution it was predicted that Pre-Cambrian formations should contain simpler fossil forms. Unfortunately, for a long time Pre-Cambrian rocks did not yield any indication of prior life. But in 1957 fossil traces of simple form were found in Pre-Cambrian formations in Australia. Once the identification was made, other finds followed. The theory of evolution had once again survived examination by some of its most demanding critics.

Another aspect of scientific work is the attempt to discover some basic structure or function that is known to exist but whose nature is unknown. A good example of scientific investigation here concerns the search for the structure of DNA molecules. Discovering ways of holding the molecules and taking radiographic pictures, interpreting the pictures in terms of recursive molecular structure, and figuring out the particular helical form of those structures is one of the most important scientific discoveries of the

20th century. Rosalind Franklin's spectrographic pictures and interpretations and James Watson and Francis Crick's models gave us the double helix and laid the foundation for much modern biological and medical research.

The most exciting and confusing—but rarest—of scientific activities is the creation of new principles or conceptual schemata. These successful revolutions are not overnight affairs. The general pattern of revolution often begins with mounting evidence that a current schema is not adequate. However, the inadequacy of a conceptual schema in and of itself does not necessarily produce a change. Rather, a revolution has to wait for someone to create a new conceptual schema. It is impossible to operate effectively without some framework. Better a rickety, worn-out conceptual schema than none at all.

To summarize, the work of a basic scientist is to organize natural processes through conceptual schemata and to collect data that test and give depth and detail to the schema. The basic scientist's primary task is to develop the concepts or principles that become more and more abstract as the science develops.

2. Applied scientists. In contrast, the applied scientist is more closely identified with the *application* of principles and concepts to a specific and generally limited problem. The forms of application are so varied that no neat summary is possible. However, a couple of examples may illuminate the applied scientist's role. Among the most visible of current applications are those in the field of rocketry. To be sure, there are some new scientific problems involved, and there has been some scientific "fallout" (that is, knowledge acquired that is incidental to the main objective), but the greatest problems have involved applications of previously existing principles and concepts. Consider, for instance, the velocity a satellite must achieve in order to go into orbit. What is involved? The gravitational field and the composition of the atmosphere are two major components of the problem. Each of these components has been the subject of prolonged scientific effort. In this situation the engineering effort that actually put a satellite into orbit used principles understood long before the attempt was contemplated. The development of the scientific principles concerning gravitation could, without too much distortion, be attributed to Newton. A more detailed knowledge of gravity resulting from "mopping up" gave engineers the approximations necessary. An appreciation of the composition and structure of the atmosphere goes back at least to Evangelista Torricelli, a student of Galileo. Torricelli's concept of a "sea of air" (about 1643) was critical to the development of orbiting satellites. This conceptual schema was developed in detail by succeeding generations of scientists.

Engineers attempting to design rocketry equipment face problems of fantastic complexity. The difference between their situation and that of basic scientists is that the problem is more specific (to put a satellite in orbit). The general concepts and details subsumed under these principles were previously available. The engineers asked "How does one use these principles to design a system that accomplishes the objective?"

A parallel example is the development of polio vaccine. Here again, the basic notion—that many diseases are the product of specific viral or bacterial agents—had long been known. In 1892, the Russian botanist D. J. Ivanovski laid the experimental foundation for research on viruses by transmitting the mosaic disease of tobacco using the cell-free juice of affected plants. By 1908, methods of transmitting poliomyelitis virus to rhesus monkeys had been developed. Work on the problem of propagating poliomyelitis virus culminated with the critical discoveries of J. F. Enders and his associates in 1949. They managed to isolate and maintain cultures of the virus in tissue taken from human embryonic organs. These historical examples illustrate the sort of scientific background that often lies behind a well-publicized application. In addition to the basic knowledge about viruses, the principles of immunity were at least partially understood, and detailed knowledge of anatomy was available. The problem was how to use these principles to generate a solution for the particular problem. A very extensive amount of research in both basic and applied science was necessary before there was any possibility of developing a polio vaccine.

3. Practitioners. The third role is that of the practitioners. Their work involves immediate contact with the materials, the patient, or the product. Thinking about general principles plays very little direct part in their day-to-day functions. Their role is to use, in a single specific case, a solution generated by the applied scientist. For example, in the case of rocketry, a practitioner (here an electrician) may be involved in positioning the instrumentation that monitors the actual amount of thrust developed at a particular stage. Another practitioner may prepare the explosive bolts that separate burned-out sections of the rocket from the payload. A medical practitioner could be concerned with the recognition of symptoms, the proper administration of antibiotics, or another specific treatment using solutions previously generated by the applied scientist.

In addition to the somewhat specific differences noted above, the three roles can be differentiated along some general lines. The number of individuals involved in each of these roles varies greatly. The number of basic scientists is small compared with the number of applied scientists, while the number of practitioners vastly outnumbers both other groups combined. The level of abstraction required varies greatly from role to role; the scientists' theories are often far removed from ordinary sense data, whereas practitioners deal with the here and now.

Generally, there is a time factor associated with these roles. Practitioners are faced with an immediate, specific problem requiring an immediate, specific solution. In most instances they can estimate quite accurately the time necessary to solve the problem. Their problems often have known solutions; their primary task is to use them. If there is no known solution, they compromise their results and do not progress toward a general solution of the problem. Applied scientists ordinarily work over a longer time span on a single problem than do practitioners. Their specific problem has not been solved before, so time limits are harder to apply. They may or may not have a programmed deadline, but they are not generally required to devise an immediate solution. Basic scientists may be harried and driven by internal desires and interests, but there is literally no way they can promise solutions on a time schedule. In some instances they do not even know specifically where an interesting observation may lead, what the solution will look like, or even whether they will arrive at one. Kepler took over 30 years to produce the three laws of planetary motion; Darwin's publication of *Origin of Species* came more than 20 years after his voyage on H.M.S. *Beagle;* and Marie and Pierre Curie literally spent their lives investigating radioactivity.

A final difference in our three groups involves their source of livelihood. Practitioners ordinarily receive their salary or fees directly for their finished products, and they are in relatively direct contact with the consumer. Applied scientists' source of income varies considerably. They may be independent consultants, employees of a firm that requires their talent in solving problems, or faculty members in a university. With the exception of faculty members, their income is directly allied to their problem area. Faculty members' time may be divided between the educational institution and outside employment, or they may work exclusively in a university setting, attacking problems simply because of their intrinsic interest.

Today in the U.S., the overwhelming majority of basic scientists is associated with universities. Financial support may be derived from faculty positions or from grants (ordinarily from the federal government) that support their research efforts. Most often their financial support has not been directly related to solving problems for consumers. Disturbing changes in the relation of scientists to consumer problems have occurred in recent years, because there are growing pressures to work on immediate problems rather than on basic ones. Evidence suggests that the greatest scientific advances are made when scientists' theoretical problems dictate their research. Historically, scientists have been the last to be supported financially for their particular work. In earlier times, some scientists (such as Benjamin Franklin, Francis Galton, and Charles Darwin) were wealthy amateurs. Others earned primary support from work not directly related to their scientific interests. Leonardo da Vinci's pathetic letter to Lodovico Sforza asking for employment emphasized his ability to contrive instruments of war. Johannes Kepler served as court astrologer and often had to plead for payment of his back wages. Other scientists have been supported from sources widely divergent from their work. Francis Bacon was a lawyer and became Lord Chancellor of England. (The appointment terminated with his conviction for accepting a bribe. The result was that he spent the remainder of his life writing. For once, the law achieved a better reform than could be expected.) Antoine Lavoisier, the great French chemist, became farmer-general of taxes and, as such, paid the government a fixed fee, for which he was allowed to collect the state revenue. This profitable undertaking supported his scientific work but was cut short by the French Revolution; his scientific work ended at the guillotine.

In summary, we have seen some of the distinguishing characteristics of three science-associated groups. Overlapping in some functions, attitudes, problems, and sources of employment is acknowledged, yet there are sufficient differences for most individuals involved to be identified and to identify themselves with one of these groups. The primary difference among the three groups is in their attitudes toward what is important in their work and the goals they hope to achieve.

Science and Pseudoscience

An introduction to science would be incomplete without some mention of Sunday-supplement science or pseudoscience. These enterprises are often fascinating and amusing, but they are ordinarily different from acceptable scientific research. We must face the fact that the confused, misguided, or dishonest cannot always be distinguished from those who merely have new and as yet unaccepted views. Pseudoscientists are well aware that many scientific innovations have been met with disbelief, scorn, hostility, and ridicule from more orthodox groups. Since pseudoscientists are often met with disbelief, scorn, hostility, and ridicule, they conclude that they must be innovators (Red-Queen logic again). Instances of rejection followed by overwhelming acceptance suggest that we should evaluate presentations carefully. These same instances provide some solace for the pseudoscientist. How then do we improve our odds for identifying the pseudoscientist while minimizing the risk of dismissing the merely eccentric or innovative?

The technique that has evolved to handle this problem is to allow both pseudoscientists and creative antiestablishment scientists to present their work. If the work has merit, it may gain followers among those who happen upon it. Ideally, a good creative idea will then snowball until it becomes a major scientific force, while the pseudoscientists' work will remain on the fringes. It should be noted, however, that most new ideas, whether from scientists or pseudoscientists, wind up under the heading of "file and forget."

Before we consider specific cases, let's recall two earlier statements. First, science is defined in terms of *how* and *why* we know something, not *what* we know. Second, scientific statements are judged by a jury of peers who are presumed to have substantial knowledge of data and experimental methodology. We should point out that a scientist essentially following scientific methodology can put forth views that are rejected by the scientific community even after they have had a hearing. Under these circumstances, observers, if they are interested in the truth, are in the predicament of having to carefully evaluate the evidence on both sides and form their own opinion. We think that it is safe to say, however, that most positions that have been reviewed and rejected by the scientific community are not as sound as the prevailing views.

We will consider some examples of varying degrees of nearness to science, ranging from obviously absurd beliefs to presumably well-intended but as yet unacceptable concepts. The issue of whether the work is well intentioned or not is important to us. We are most annoyed by those who pretend to do science but whose goal is to fool a gullible public and get rich from it. Perhaps we are annoyed because we would like to be able to make the millions that they do. But *Chariots of the Gods?* by Erich von Daniken, which purports to be a scientific theory of ancient extraterrestrial beings, is filled with false statements, as is such tripe as the books on the Amityville horror and the Bermuda Triangle. These works make no attempt to verify statements just so long as they sell. Most people generally believe what they read. We accept attempts to be creative and innovative as a virtue even when ideas are half-baked. We can find much better use of time and money than in supporting con artists or attempting to refute them; however, we cannot always discriminate the con artist and the pseudoscientist from the unorthodox scientist.

By the beginning of the 16th century it had become less dangerous to believe in the earth as a sphere. Yet the belief in a flat earth still persists among some people in the United States. A rather strange fundamentalist religious group holds this view and defends it with what it considers evidence. According to "flat-earthers," the earth is flat and disk-shaped with the North Pole in the center and the South Pole ringing the outer edge. Sailing around the earth is merely describing a circle on this flat surface. The earth is motionless as well as flat, and the sun is 32 miles across and about 3,000 miles away. The fact that a person who jumps straight into the air for one second doesn't come down 190-odd miles away was considered evidence that the earth is motionless. A picture taken across 12 miles of lake or a view of the opposite shore through binoculars was supposed to indicate that the surface is a plane. The "evidence" supplied by the flat-earthers was obviously highly selected to defend their position and did not take into account a myriad of facts that would refute it.

The second and slightly more sophisticated example of pseudoscience starts in dogma and ends in disaster. This one, mentioned earlier, is popularly known as Lysenkoism and involves the field of genetics. At an earlier time there were many substantial scientists, such as Chevalier de Lamarck and Darwin, who either accepted or toyed with the idea of the inheritance of acquired characteristics. According to this theory, an animal such as an elephant originally had to reach up for food. This reaching stretched the trunk, and future generations inherited this elongated trunk. The rise of modern genetics and overwhelming experimental evidence deflated this idea by the early 1900s. Enter the Marxist dogma. According to Marxist fundamentalists, everything depends on the environment. The development of an animal or a plant is determined by its surroundings. Accepting this dogma, Trofim Lysenko concluded that plants grown under a particular set of conditions would acquire characteristics determined by those conditions, and the characteristics would then be transmitted to their descendants. That great scholar and scientist Joseph Stalin agreed and confirmed his position with brute force. Western observers visited Lysenko's farms, where he claimed to have achieved astounding results. These observers were overwhelmingly impressed by Lysenko's lack of understanding of controlled observation and his substitution of harangues for evidence. Showing patience beyond belief, they repeated Lysenko's experiments in other countries; the results were decisively negative. Despite this evidence, Lysenko held sway over virtually all Soviet genetics for about 20 years.

The third example perhaps brings us to the fringes of science. This is the supposed phenomenon of extrasensory perception (ESP). The research to be considered is primarily that of Dr. J. B. Rhine. Rhine was certainly not in the same category as the flat-earthers or Lysenkoists. He made a sustained effort to establish ESP as a scientific phenomenon, but after approximately 40 years of research and writing, his views and the ESP phenomenon failed to be accepted by most scientists. His failure is primarily one of evidence and integration. The great majority of independent tests do not produce results supporting his concepts. In addition, his methods of analyzing data were simply not accepted by most other scientists. Probably his most fatal flaw was the inconsistency of his results. His subjects had a lamentable tendency to return to the statistical mean. Crapshooters are familiar with this sort of problem. Simply stated, it means that sometimes you will have hot streaks even with honest dice—for example, when 7 comes up five times in a row. But over a long period the shooter drifts towards the statistical mean, with 7 coming up an average of approximately once every six throws. If one selects only the hot streaks, it looks as though the shooter is controlling the dice. Data selection seemed to be basic to Rhine's interpretation. This is not selecting what kind of data to investigate, as other scientists do, but rather selecting "good" data from a pool that one collects. This second kind of selection clearly must be generally unacceptable to science. Even if Rhine and his colleagues were dealing with a real phenomenon, all they did for 40 years was to attempt to prove its existence. Scientists ordinarily integrate the phenomena they establish into a conceptual schema; Rhine's group failed to do this. Another item created difficulty for Rhine. This is the notion that whatever is being investigated is extrasensory; in other words, information is being transmitted by means that cannot be measured with any known instruments. Such an assumption is contrary to current scientific theories. This assumption, bizarre as it may seem, does not by itself rule out consideration of ESP as part of science. The lack of acceptable evidence and interpretation of reliable phenomena are the fatal flaws.

Despite the obvious differences in the three examples, they have some common problems that leave them outside the accepted realm of science. The central problem is one of the quality of supporting evidence and the conceptual schema. In Chapter 4 we will discuss the nature of scientific evidence. For the moment, consider evidence in the same sense that a jury might. In a trial the burden of proof is on the party bringing the suit or making the charge; the charge holds a position similar to the new proposal. In science the burden of proof is, and rightly so, on the one challenging an accepted belief or attempting to establish a new conceptual schema. There is no requirement in science that we must attempt to disprove any particular assertion. The flat-earthers were completely unconvincing because their evidence failed on all counts. The assertions they made were highly selective, they violated the rules regarding what constitutes evidence, and, most important, their concepts simply did not take into consideration other existing evidence. Lysenko failed in that his evidence could not be supported by external sources; he failed to have even a rudimentary understanding of how evidence is collected. He also failed to produce concepts that accounted for other well-substantiated observations. Lysenko was probably dishonest and was assuredly a political hack, although his character is irrelevant as far as the scientific worth of his work is concerned. His concepts and evidence can be judged independently of him. Finally, Rhine's theory failed to be accepted primarily because of the inconsistency of the evidence.

Our major goal in this chapter has been to demonstrate that, despite their agreement on the importance of science, scientists and lay persons have very different ideas about what science is and how it operates. The critical difference is that many nonscientists think of science as a collection of facts leading to practical ends, whereas scientists think of science as a set of methods and conceptual schemata leading to an understanding of natural processes.

3

Science, Dogma, Mysticism, and Superstition

Something about scientific attitudes makes them seem significantly different from the other attitudes that have dominated cultures and subcultures. But what is it that makes science unique? This problem has been considered by scientists and philosophers for centuries, and there still are disagreements. However, the disagreements are relatively small compared with the areas of agreement and are generally related to items within science rather than the general attitudes or methods of scientists.

The person who plays the game of science has a clear idea of what is important in deciding whether a statement is to be accepted. The criteria for accepting or rejecting any statement are the logic of the arguments within a statement and the arguments' relationship to the data. Scientists may make mistakes or may be misled by false premises or emotions, as anyone may, but the rules for their decisions are well understood. The rules have evolved through centuries of effort by philosophers and scientists who have thought about and tested many possible approaches to science.

In this chapter we will compare ways of attempting to understand and/or interpret the world. We repeat the quotation from Bertrand Russell: "It is not what the man of science believes that distinguishes him, but *how* and *why* he believes it." The *how* and *why* of other belief systems make an interesting contrast to those in science.

Scientists, while playing the scientific game, make no decisions on the basis of faith (they may make assumptions, but these remain open to question); no decisions on the basis of power; no decisions on the basis of monetary rewards; and no decisions on the basis of self-protection. Scientists have to be intellectually honest. The bases of their decisions must be observed events and an attempt at their interpretation. It is true that on some occasions scientists deny seemingly obvious and necessary conclusions about observed events and their interpretations because of preconceptions, but in doing so they are not acting as scientists. If they make a discovery that contradicts what they "know" from their religious, economic, or other beliefs and summarily reject this discovery, they are not playing the scientific game according to the rules.

Because of the importance of data in determining the beliefs of scientists, a tendency has developed to equate data and science. This misunderstanding was discussed in Chapter 2. Scientists' decisions and conclusions are validated by data, but the data per se are not the science. Scientists gather data and study them, but to make the whole enterprise scientific they must also organize and interpret the data. A more specific discussion of how data are organized and interpreted appears later in the book.

Two people can hold the same beliefs and have the same knowledge. One may be a scientist and the other not. For example, if all of the scientific information known today were to be studied and accepted with no new work done, there would be no science. If in answer to the question "How does energy relate to matter?" you read that $E = mc^2$ and learn that "energy in joules equals the mass converted in kilograms times the speed of light in meters per second squared," you are not necessarily playing the scientific game. A statement might look like science and sound like science, but it is not science if the only reason you believe it is that you have read it or Einstein has said it. Having knowledge is not what makes a scientist. It is the *method* of attainment of knowledge that determines whether one is playing the game according to the rules.

Aristotle was a brilliant man; he is remembered primarily as a philosopher and logician, but he was also a scientist. He observed events and thought about them. He made many statements related to these observations. The truth or falsity of these statements is not critical in determining their scientific status; only their method of attainment is critical. The acceptance of these statements during the late Middle Ages by the Scholastics was not scientific, because for them truth depended solely on the fact that Aristotle had made the statements. They did not justify their belief by reference to data.

Science versus Dogma

For those playing the game of science, the method by which they ascertain what they believe is crucial. They have to evaluate data and arguments and decide for themselves on their validity. Scientists communicating to others have the task of convincing the hearers of the validity of their statements in terms of the data and their explanations of the data. They are not playing the game correctly if they win support by the strength of their personality or prestige. In the short run scientists may follow another on the basis of prestige or charisma, but this cannot continue indefinitely. It is up to the hearers to evaluate what the scientists say rather than accept it because they say it. And the culmination of this enterprise is the determination of whether the explanations account for the data observed. The relationship between scientific explanations and data helps to ensure the integrity of scientists. If their findings are important, they can be sure that their theories and data will be examined critically and new evidence attained.

It is the system of data-based explanation that distinguishes science from dogma. Scientists have both the right and the responsibility to decide for themselves, on the basis of the evidence at hand, the best explanation of a set of phenomena. They also have the right and the responsibility to investigate thoroughly the bases of their beliefs. They cannot accept statements unsupported by data. On the other hand, dogma (religious, economic, political, social, or any other kind) depends on pronouncements by established authorities (for example, the dogma that the earth was created in 4004 B.C.). The goal of students learning a dogma is to accept the pronouncements as they are given. If the students disagree with the dogma, they are not playing the game of dogmatism correctly. They have to search their souls until they accept it or be considered outcasts and suffer the consequences. It would not matter if they could present strong arguments in support of their personal beliefs. In dogma, arguments and facts are forced to coincide with the dogma. Students cannot accept statements that do not agree with the dogma. (Continuing the previous example, they must reject the existence of prehistoric people around 10,000 B.C.)

One way of contrasting science and dogma is to say that a scientist accepts facts as given and belief systems as tentative, whereas a dogmatist accepts the belief systems as given—facts are irrelevant.

It is possible for dogma to exist in the realm that is normally thought to be scientific. For example, if physicists accept the theory that a man who was accelerated away from and back to his twin would be younger than his twin when he returned (the twin paradox) because Einstein said so (rather than by understanding the arguments and data—if any—that confirm it), they are acting dogmatically. In fact, in *Physics Today*, Mendel Sachs argued that there are, in fact, no data supporting the theory and that the arguments of the twin paradox are fallacious. Reader response treated him like a heretic rather than a physicist with another opinion. It is difficult not to be dogmatic about one's beliefs, whether they are justified or not. The theoretical statements of many scientists have been made dogma by their followers. That is, their followers accepted these scientists' statements as absolute truths. Such unfortunate scientists include Aristotle, Ptolemy, Newton, Lamarck, and Freud.

Science, Mysticism, and Superstition

Chapter 2 presented a discussion of attitudes and beliefs about science, both accurate and inaccurate. In this chapter we are interested in briefly exploring avenues of knowledge that do not meet conditions generally deemed scientific. We are both puzzled by the fact that in recent years a very large number of Americans have adopted or maintained a belief in many variants of supernatural, mystical, and magical powers and sources of information. We haven't had many heroes in the United States recently, but the void seems to be filled in part by gurus, seers, and magicians. What do they have to offer? What evidence supports the tales of mysterious disappearances in the Bermuda Triangle, or persuades 50% of the people of the United States to believe in exorcism, or convinces the millions upon millions who follow horoscopes?

Science is usually defined primarily by its method of attaining information. The beliefs held by scientists, insofar as they are playing the game, are part of a coherent system of ideas supported by observations. However, scientists usually have another aspect of their belief system that they apply to their theories, and that is a general belief that rational and natural concepts will ultimately be adequate to account for whatever data need to be explained. This does not mean that the scientist has explanations for all observations or that we necessarily know about every kind of power and energy but rather that, when an explanation is found or offered, it will be one based on the assumption of the regularity and the rationality of the underlying processes that caused the phenomena. There is one possible exception to this position. Quantum theory, which seems to be descriptively consistent, does not present a clear, coherent set of underlying processes from which the descriptions are derived. Rather, the observations are precisely accounted for by objects that, according to some previously specified probability, may or may not appear at the time of the observations.

The assumption of rational processes has led to essentially all scientific discoveries and technological inventions, including electricity, neural action, magnetism, radioactive decay, television, airplanes, bifocals, antidepressant drugs, and kidney transplants. Yet over half the populace believes in strange and nonnatural powers and spirits as explanations for unusual and usually undocumented events.

Earlier we quoted Bertrand Russell's statement that scientists are distinguished by the how and why of their beliefs. The same could be said of mystics. However, the how and why are quite different for the two groups. There is a superficial similarity in some of the experiences of scientists and mystics. These are the experiences in which there is a sudden inspiration or insight into a problem, which is immediately seen in a new light or completely restructured. In spite of the similarity, the process emphasizes the difference between scientist and mystic. The mystic attributes the experience to some unknown or supernatural event. The scientist does not. The most important distinction is what happens after the insight. The mystic is content, maybe joyful, in proclaiming the experiencing of eternal truth. In contrast, the scientist is anxious to put the inspiration or insight to the test of data. Sad to report, many insights wilt in the glare of evidence. The mystic doesn't take that risk.

As we use the term in this book, *mysticism* includes the belief that knowledge can be gained through nonsensory means. Just how this knowledge is gained is often left vague. For example, the French philosopher Henri Bergson felt that absolute knowledge is possible only through intuition. Just what he meant by intuition is really not quite clear, but it is obviously not based on ordinary sensory information. As W. T. Jones saw the situation:

> But are we not then face to face with . . . the paradox of trying to philosophize about the ineffable, and by means of a conceptual analysis on whose inadequacy the author proudly insists? As a matter of fact, in Bergson's writing metaphor and imagery tend to take the place of argument and proof . . . When we take it as a theory and ask for evidence, we are reminded that evidence is only a fiction created by intellect in its own image.

There are many different kinds of mystics; Bergson is used only as an example. Not all mystics would take such a cavalier attitude toward evidence as Bergson did. For example, Rhine and some other believers in ESP tried to establish scientific evidence for their beliefs, with no real success to date. In spite of their general scientific attitudes, Rhine's followers were willing to accept rather weak evidence and explain away negative evidence—more on this shortly. Their attitudes toward evidence were generally different from those of most scientists.

Although quite similar in their approaches to "understanding" events (or movements), superstition and mysticism differ with respect to their subject matter. Superstition is the (ordinarily strong) belief in mysterious powers that control our lives and/or the physical universe.

A relatively harmless example of a superstition is the football coach's wearing of his "lucky" hat. This works well in some seasons, but without the big, fast, mean linemen, shifty backs, and so on, the hat seems to lose its effectiveness. Interment of the failed fetish and relief of its wearer from his character-building endeavor may well follow. The poker player who walks around the chair three times to end a streak of bad luck would be better advised to stop drawing to inside straights and bobtailed flushes. Gambling casinos, which have people using every lucky charm imaginable, do very nicely winning on mathematically computed odds.

It is not surprising that people continue to gamble—they win sometimes. Scientists have shown that both humans and nonhumans rewarded at random times continue to respond for a long time. Likewise, human superstitious behavior is usually not causally related to any apparent payoffs. If people do something that happens to be followed by a relatively unexpected reward, they may repeat the behavior. If they repeat the behavior often enough, it is likely to be rewarded again, simply by chance.

There is a way superstitions can affect outcomes. This is called a self-fulfilling prophesy. Under these circumstances people change their behavior because of a belief in a superstition. For example, baseball players may not play as hard if they don't have the right socks on, because they "know" it won't do any good. Or people may become very tense after a black cat crosses their path, with the result that they overreact in a normal driving situation and cause an accident.

Another superstition—one that has a more serious implication—is the belief that illnesses can be cured by chants or charms. This superstition may produce disastrous results for everyone concerned, except the purveyors of chants and charms.

Characteristics of Mysticism and Superstition

There are a number of ways in which mysticism and superstition differ from science. First and foremost is the origin, kind, and quality of evidence that is acceptable. Martin Gardner wrote an amusing review of an attempt to demonstrate ESP by machine. He phrased some of his comments in terms of Joseph Heller's delightful classic, *Catch-22.*[1] (If you haven't read it, there's a bunion on your academic standing.)

> Catch 22: Skepticism destroys the subtle operation of psi. It is a catch unique to parapsychology. In other sciences failure by a doubting scientist to replicate an experiment is counted as disconfirming evidence. Because psi powers are said to be adversely influenced by doubt, however, parapsychologists are not impressed by replication failures unless they are obtained by sheep.[2] In this case no goat[2] was present, so that Targ turned to Catch 23.
> Catch 23 asserts that psi powers are negatively influenced by complexity. As Rhine once phrased it, '. . . elaborate precautions take their toll. Experimenters who have worked long in this field have observed that the scoring rate is hampered as the experiment is made complicated, heavy, and slow-moving. Precautionary measures are usually distracting in themselves.' Catch 23 achieves a truly remarkable result. It makes it impossible to establish psi powers by tests that are convincing to the goats who are the vast majority of

[1]Catch-22 employs the principle of heads we win, tails you lose.
[2]Note that a goat is one who doesn't believe in ESP; sheep do.

> professional psychologists. As long as testing is informal and under sloppy controls, you get results. If you tighten controls, the experiment inevitably gets complicated and scores fall.[3]

The story of the failure of Targ's expensive ESP experiments is emblematic of what has happened numerous other times in ESP research. High-scoring subjects are first identified by loosely controlled screening; then, as testing proceeds under better (that is, more complex) controls, their psi powers mysteriously hide. In addition to Catch-22 and Catch-23, parapsychologists have a string of other good ones. Catch-24 says that, for reasons nobody understands, high-scoring subjects tend to lose their powers.

The quality of evidence acceptable to mystics and the superstitious differs greatly from the quality of acceptable evidence in science. A recent article in *Reader's Digest* told the story of a woman (a sheep) who, after a long search and many startling coincidences, found her long-lost cat. This was considered evidence for ESP. Scientists would consider this an old wives' tale without serious verification, controls, or other precautions that separate the subjective from the objective.

Seances, seers, levitators, mind readers, the Bermuda Triangle, the Loch Ness monster—whenever *any* of these and other such "supernatural" phenomena are carefully investigated, the mystery disappears. Evidence is uncovered of cheating or distortion of facts; natural explanations of the remaining phenomena are found. These mysteries are always much more impressive secondhand.

In fact, when someone reports such phenomena it has often been witnessed by what we have heard described as a "foaf" (friend of a friend).

A second characteristic—almost a hallmark—of mysticism and superstition is a high degree of vagueness together with simplistic statements that could fit almost anyone or any situation. The following, for example, was taken from a newspaper column by a highly touted astrologer and seer.

> Thursday, October 21: If your birthday is today, it will find you designing a major move to get out of past and present limitations. Direct your main effort all year to the job of scraping together enough resources to finance yourself and reduce your dependence on mate or business associates. Relationships thrive as you meet people of similar tastes. Today's natives adopt a variety of vocations, support unconventional causes. Those born this year will face a special challenge, but respond to it with clever ideas and sophisticated methods.

So you're "designing a major move." Who doesn't have at least vague ideas of the big break? All this is really saying is that people dream of a better future. "Today's natives adopt a variety of vocations." This has the

[3]From "Mathematical Games," by Martin Gardner, *Scientific American,* October 1975, 233(4). Reprinted by permission.

profound implication that somehow people differ. Given vague, broad statements such as these, you can't miss.

A third characteristic of mysticism and superstition is that the beliefs are strongly held, often in the face of substantial contrary evidence.

Belief in astrology has survived and thrives in spite of a lack of positive evidence and in the face of both contrary evidence and analysis showing that it is not even remotely likely to be valid.

Belief in ESP continues, even though after over 50 years of research there is still no acceptable evidence for its existence. Remember Gardner's comments. We have conducted a number of ESP demonstrations in class, sometimes using sheep and sometimes both sheep and goats. The results, without exception, have given data that are reasonably accounted for by chance. This is hardly surprising. What is frustrating is that the true believers remain undaunted.

For over a thousand years there have been innumerable predictions, generally by religious groups, that the end of the world is to occur on a particular date. People may gather in their homes or with a leader of the cult. The appointed time comes; nothing happens. Some desert, but others find alibis: the data had been calculated wrong, their meeting in faith had brought a reprieve, and so on.

Although they tend to be more vague about the exact dates, a variety of radio or TV evangelists say this same sort of thing any day of the week. James Watt, at one time U.S. secretary of the interior, was opposed to conservation because THE END was near. Maybe it will happen when the New Orleans Saints win the super bowl.

Belief in the creation story and consequent opposition to the theory of evolution continue to be very strongly held. Recent events in courts and legislatures of many states (over 20) have been directed toward downgrading or suppressing the teaching of evolution or giving the creation story equal billing in biology classes. Again, belief triumphs over evidence.

Consider the story of Noah's ark. It has been estimated that there are approximately 350,000 species of plants and 1,150,000 species of animals. Some of these could survive underwater; others couldn't, depending on whether the water was salt, brackish, or fresh. It seems likely that well over one million plant and animal species couldn't. Collecting species at the rate of one pair every five minutes (two and a half minutes each for males and females) would yield 864 species collected per 24-hour period (no coffee breaks, meals, sleep, or anything else considered). At this rate collecting 1,000,000 species would require 1157 days and nights to complete. Even if the plants are excluded (they aren't mentioned directly but probably were included as food—Genesis 6:21), it would still take 878 days, or nearly two and a half years, to load the ark. Just how you collect two polar bears in five minutes starting from the Middle East escapes us at the moment.

We have considered only species. Genesis has been translated as "kind" and "sort"; presumably this would cover the subspecies, which would multiply the numbers considerably. Further, the clean beasts and fowl were to be taken by sevens (Genesis 7:2–3). In any case the number is fantastically large. Loading even 1,200,000 pairs in seven days (Genesis 7:4 & 7:10) would require putting them on board at the rate of 115 pairs per minute, or 1.916 pairs per second. (It really isn't clear whether they were to be collected or whether they reported on their own.) To cover the last 100 feet approaching the ark in 0.52 second would require a speed of approximately 130.8 miles per hour. The snails, the tortoises, and the 600-year-old Noah would really be in high gear.

Questions could also be raised concerning food, water, and space required, life expectancies, possibilities of sterility, and many other things—but enough. It seems fair to say that in physical terms, *as we know them,* the situation is impossible. Here we reach the critical difference between science and superstition. Believers in the literal truth of the Bible could concede the evidence of our numbers (which are approximations, probably on the low side) and still present their case on the basis that the ark was loaded by miracles in which ordinary physical laws are not applicable. How do they know? By belief. The scientist would point to evidence and logic. The issue is drawn: do you accept belief or evidence and logic?

Sources of Mysticism and Superstition

What processes in humans lead to a belief in mysticism and superstition? We can't be sure of all of the factors. We can, however, suggest factors that we do know operate in some cases. The world is a confusing place; our powers to control it are very limited; we are fearful of the unknown and uncontrollable. Although these principles would apply equally to modern and ancient man, an example from earlier, simpler times may be useful. Consider Joe Neanderthal as he views a thunderstorm. Lightning flashes, thunder rolls, winds howl, and rain falls in torrents. Here indeed is a confusing, uncontrollable, fear-provoking phenomenon. How is this awesome event to be understood? Rocks, dirt, and trees don't move of their own accord, but animals both move and make sounds. Perhaps the storm is some strange, huge beast. Since beasts were and are considered in many primitive societies to have human characteristics, it is only a short step to attributing human characteristics to the storm. But something so powerful and dreadful must indeed be an extraordinary creature capable of a mighty wrath.

Things have not changed all that much. Benjamin Franklin was denounced from the pulpit for preventing a righteous God from inflicting just punishment on the wicked by inventing the lightning rod. Just how an omnipotent God would be stopped by a lightning rod is a logical puzzle, to say the least. Recently, after a tornado, survivors proclaimed on TV that they had been spared by the grace of God. Their neighbors one block away must have been exceedingly and uniformly wicked, since their homes were devastated without exception, and many persons died.

One thing stands out from this discussion. Humans want answers to the unknown and fearful. If answers are not available or understood, then answers will be invented. Probably necessary in a prescientific era, this proneness to invent answers is lamentable today.

A second source of mysticism and superstition is based on learning. In a wide variety of animals, such as pigeons, rats, and humans, we find a tendency to repeat behaviors that have preceded a reward. Further, we know that the reward does not have to occur on every occasion. In fact, if the reward follows only part of the time, the response may be more vigorous and may continue longer after all rewards have ceased. Consider our earlier example of the football coach. He wears the hat and wins. Reward! It does not require many instances of these joint events to ensure the behavior. Interestingly enough, if he wore the hat during an all-victorious season followed by an all-losing season, the hat wouldn't last too long during the second season. On the other hand, if his first season had a mixture of wins and losses, he would continue his hat wearing longer into the second season.

Horoscopes work the same way. The individual gets a prediction couched in language sufficiently vague to cover a multitude of possible events. One of the possible events occurs. Reward! The tale follows the dreary course of hat wearing.

A strong belief without evidence often leads to distortion of the evidence. Gardner reported that "Walter J. Levy, Jr., Director of J. B. Rhine's Institute for Parapsychology, resigned after it was found that he had been tinkering with the apparatus to improve scores." Many people—in particular, the magician James Randi—have investigated Uri Geller, the Israeli mystic and psychic, and shown that Geller does not perform willingly around magicians who can spot his tricks; when care has been taken to eliminate tricks, his "psychic powers" don't work. According to Leon Jaroff, senior editor of *Time* magazine, "After reading Randi's logical and rational explanations of Uri's tricks, only the most fanatic and gullible followers (alas there are many) will continue to believe in Geller and in the psychic phenomena he claims to effect."

Other mystics have the same success following investigation. Jeanne Dixon is known for her prophecy, but no one seems to remember her misses. At different times she has predicted that Nixon, not Kennedy, would win the presidency in 1960; that Walter Reuther would win in 1964 (we don't remember that result); that Ford would win in 1976 (he didn't do badly—he came in second); that Russia would be the first nation to put a man on the moon (too bad no one told Neil Armstrong and Edward Aldrin, Jr.). She has made other equally accurate predictions. Her revelations seem to be as accurate as anyone's who has a small amount of general information and is willing to go out on a limb. In fact, for most of her predictions she doesn't even go out on a limb. Indeed, she presents general philosophical statements that are untestable.

The domain of the supernatural is fraught with charlatans making money off a gullible public. The will to believe, coupled with strong feelings that nice, friendly, open, sincere people would never lie or misrepresent the facts, can win many converts who think they have seen mystical, magical phenomena. Watch a good stage magician whom you know is using misdirection and trickery; then judge what deception may get through as real magic. Be aware of the fact that eyewitnesses in courtrooms are notorious for their misrepresentation of facts even though they are certain that they are reporting correctly. Their recollections are affected by the discussions they have had after the event as well as by their expectancies beforehand. Read a book such as Randi's on Geller or Christopher's on the occult (see the Bibliography). Check predictions made by ESPers, astrologers, or psychics. People are creative, insightful, smart, talented, personable, happy, knowledgeable, and many other things. The world is variable, complex, strange, interesting, changeable, and many other things. But there is reason to believe that all of these exciting and complex phenomena are consistent with and part of natural processes and thus investigative by scientific methods and concepts. When you read or hear something that seems to be out of the ordinary, ask yourself: "How does the author know? Is supporting evidence presented for the claim? Are there other, more straightforward explanations of the report? Is there reason to expect trickery or misinformation? Who stands to profit?"

Science doesn't have all of the answers, and it never will. However, some reality testing, a willingness to disbelieve on the basis of contradictory evidence, and an awareness that a belief is open to challenge are all part of the game of science.

4 Rules and Concepts

Basic Concepts Involved in Playing the Game

We now turn to the details of playing the game of science. What counts as data and what counts as an explanation of data? These are questions that have controversial answers. Scientists do similar things, but they don't always evaluate what they do in the same way. One major reason for this disagreement is that one cannot in any satisfactory way separate data from explanations of those data. Attempts have been made to separate them, but no formal analysis of the difference between data and theory has been satisfactory to all concerned.

Your understanding of the game of science may be aided by a brief description of observable events, data, laws, and explanations, along with a discussion of their similarities and differences.

Observable Events and Scientific Data

An event does not enter the domain of science merely by occurring. It must be recognized and recorded in some form. A datum is the recording of that event. In other words, the datum is the representation of the event by some means that is relatively permanent. The event itself occurs and then is gone forever. The datum is usually recorded in a symbolic form (words or numbers) at the time the event is observed.

Consider the following four statements.

1. There are 12 books on Professor Brown's desk right now.
2. There was a thunderstorm in Chicago last night.
3. The temperature outside is currently 28° Celsius.
4. He traveled 262 miles in his car between the last two times he filled the gas tank, and on the last occasion he added 11.8 gallons of gasoline.

As written, these statements are data if they represent actually occurring events. As data, however, they are very incomplete, because they do not accurately identify the specific events they refer to.

In addition to statements like the four just listed, science requires sentences like the following four, called laws, generalizations, or inferences.

1. Professors keep many books on their desks.
2. It rains a lot in the springtime.
3. It is warm in the afternoon.
4. His car gets approximately 22.2 miles to the gallon.

The major difference between the two sets of sentences is that the first set describes *individual* events, whereas the second set describes *classes* of events. In science, relations between classes of events are, in part, inferred from observations and recordings of individual events. The scientist collects data by observing singular events and then bases generalizations on the data. However, that an event is thought to have occurred does not mean that it is automatically accepted as basis for a datum. Most scientists believe that an event has to meet certain other basic criteria.

The first criterion is hard to formalize but is very important. The event has to be one of a type that, in principle, more than one person could describe. This criterion is known technically as *intersubjective testability*. The current position of most scientists is that no event is admissible as the basis for a scientific datum unless it is intersubjectively testable. One way of assuring intersubjective testability is to require that all recordable events must be ultimately describable in terms of tables, chairs, colors, sounds, pointer readings, weights, and so on. That is, the events have to be describable in the language of physical things (called the "physical-thing language" by some philosophers).

It is rather difficult to describe accurately what can be included as physical things or intersubjectively testable events. Informally, what is important is that the event described can be witnessed by more than one person and that the description of the event can be understood by most other people. It is interesting to note that sensations and pains do not meet this criterion; they cannot be witnessed by more than one person. But reports of sensations and pains are acceptable. In other words, the event that is acceptable as scientifically observable is someone's saying that he or she has pain, not the pain itself.

The criterion of intersubjective testability is an important one, and one that is generally thought to be necessary. It means that what you see in a dream is not a scientifically observable event, because no one else can see your dream. If you have an itch or a pain, that is not admissible directly as a scientifically observable event because no one else can feel your itch. Psychologists can accept as a scientifically observable event your report of a dream, a pain, or a wild sensation; but it is accepted as a report, not as the actual event. They can gather some supporting evidence from other sources, such as neural activity, rapid eye movements, or other physical indices. We, however, have no way of directly verifying a sensation such as a pain, nor do we have any expectation of being able to do so in the future.

A second criterion for an event to be admissible to science is the requirement of a high degree of agreement among different people on the description of the event. This criterion of *reliability* is related to but somewhat different from the criterion of intersubjective testability. Intersubjective testability requires the event to be open to public view and not entirely personal. Reliability requires the event to be described in such a way that different individuals can agree on the description. It is this description of events, not the events themselves, that scientists enter as a datum, although to simplify communication many scientists refer to the event rather than its symbolic representation as the datum.

Most events can be described so that different individuals will agree on the description. For example, one might simply say that an event occurred. Since everyone would agree that an event occurred, that description is highly reliable, but it is not very informative. Reliability has associated with it a criterion of *precision*. The more accurately and specifically an event can be reliably described, the more acceptable the description is to scientists. Not only should the event be described so that different individuals can agree on the description, but it should be described in such a manner that the event can be differentiated as much as possible from similar events along relevant dimensions. You may describe the temperature of a bowl of water as warm or cool, and other people may describe it in a similar way. Another person's verbal description "warm" describes a range of temperatures that probably is similar to but different from yours. You can make the description both more reliable and more precise by using a thermometer and describing the temperature as 25°. By describing the temperature in this manner, you differentiate the event much more precisely from other events, and different individuals can more readily agree on the accuracy of the description. Precision is one of the goals of scientific description of events.

To help improve reliability and precision as well as for other reasons, some of which are obsolete, certain philosophers and scientists introduced the concept *operational definition*. Like the other criteria, this one cannot be formally defined, but it can be discussed and used. An operational definition is the description of how one gets data. For example, an operational definition of the temperature might be "A mercury thermometer that was in the shade, over grass, 4 feet off the ground, in the open on three sides and above, gave a reading of 41° C at 2:00 P.M." The description cannot be totally complete without being extended forever, but many relevant variables are mentioned. When scientists write, they imply many things without stating them, and they ignore other things because they think they don't make any difference. "There was no fire burning nearby," "The thermometer was 10 inches long" or "The thermometer contained 1.7 grams of mercury." As you can readily see, possible descriptions of the conditions under which data are collected are inexhaustible, so no operational definition is complete. Despite its incompleteness, however, an operational definition can improve specificity in scientific communication.

In summary, there are three main criteria applied to events for judging whether they are admissible as the basis for scientific data: (1) The event, to be described by a datum, must actually occur. Data are symbolic representations of singular events. Interpretations and generalizations given to these events are not themselves data. (2) The event must, in principle, be available for public scrutiny. It does not necessarily *need* to be sensed by more than one person, but it must be of a type that *could* be sensed by more than one person. (3) The description of the event should be such that different individuals can know, as specifically as is reasonable, what the event was that is being described.

Scientific Laws

Although scientists use individual events for data, their explanations always rest with classes of events. There are many classes of events, however, within which observations are so well established that a *law* or *generalization* is accepted by almost all scientists. For example:

1. The speed of light is a constant of approximately 186,000 miles per second.
2. All cultures have a form of the family as a principal institution.
3. Oxygen is about 16 times as heavy as hydrogen.
4. The gestation period of elephants is about two years.

These laws, or classes of events, are what many people erroneously call science. They are the "facts" that people think of as the defining characteristic of science, although, as we have seen, just knowing these laws is not the most important thing in playing the game of science. Laws do not define science, but they are some of the most important products of science. They are the principles used by applied scientists and practitioners in solving specific problems.

What, then, are scientific laws? Laws are statements describing different properties that go together in the same kind of event or in certain sequences of kinds of events. They are descriptions of relatively constant relationships between certain kinds of phenomena. These descriptions may be in sentence form or in symbolic form, such as an equation. It is important to remember that laws do not refer to singular events; rather, they refer to *any* singular event that has the properties defined in the law. This last characteristic of laws is the major reason that Rhine's research didn't qualify. He had no prior criterion of which data follow the "law" of ESP and which do not.

Scientists expect all events having the necessary properties to conform to the law. As indicated before, specific measures of events lead to probabilistic statements. On this basis we would not expect any single measurement to conform to the law exactly. But scientists do expect relationships between kinds of events to be relatively consistent. It is the reproducibility of phenomena that leads one to accept a relationship as a law. The fact that any object starting from rest in free fall in a vacuum shows the approximate relationship $S=\frac{1}{2}gt^2$ leads one to call this statement a law. The law that oxygen is about 16 times as heavy as hydrogen was accepted, among other reasons, because decomposition of water regularly yields about two volumes of hydrogen to one volume of oxygen and because the oxygen produced always weighs about eight times as much as the hydrogen produced. Laws are established by the consistent repetition of relations between kinds of events, not by a singular occurrence of any succession of events.

Problems arise for scientists when an event that is expected to obey a law does not. We cannot punish the event; like the customer, the event is always right. If an event does not obey a law as it should, it is the law that is at fault, not the event. It may be, of course, that the scientists had made an error in their experimental setup, so that the event did not have the characteristics stated in the law. There may have been errors of measurement or of interpretation as well. If a law has been repeatedly confirmed, scientists are usually very reluctant to modify it; so they are likely to check carefully to see that no errors were made before they attempt such a modification.

So far, this discussion of scientific law is quite abstract; let's look at a few short examples. It is a primitive scientific law that things fall. This law means that on earth any event having the property of thingness moves in the direction of the center of the earth when support is taken away. Like all others, this law has specific applications and limitations. A child's balloon when filled with helium does not fall; it rises. So a qualifier must be made. After much experimental research, scientists concluded that "support" includes air. Air is a fluid and has weight; therefore it can support anything that for a given volume weighs less than air. Just as a boat sinks until it displaces water that weighs as much as it does and no more, a balloon will rise until it displaces air that weighs no more than the balloon. If the supporting air is removed, the balloon will fall. Note that a scientific law depends on many conditions, not just a single one. Theoretically, a scientific law holds for all events that have the right set of properties.

Continuing the same example, one might ask: "What about birds and airplanes? They don't fall all the time." The answer to this query is a little more complex, but essentially it again leads to a limiting of the conditions under which the law that things fall holds. Certain surfaces, like wings, when moving through the air set up a partial vacuum that, in effect, increases the volume of air displaced until the weight of the air displaced is comparable to the weight of the winged object. Thus birds and airplanes would fall except for the fact that in flight their wings generate a force to counteract the fall.

Another example of a law is that living organisms are composed of cells. In other words, anything that can be defined as a living organism has as its building blocks one or more relatively self-contained units known as cells. This law states that if an event has one set of properties (for example, "living") it is expected to have another set ("cellular structure"). This particular law was established as universal by a biologist, Theodor Schwann, in 1839. Like the one dealing with falling bodies, it requires elaboration and identification of limiting conditions. Certain aspects of an organism may not be cellular, because cells secrete noncellular material (hair, hormones) that remains in or attached to organisms. Also, a problem with this law may exist with respect to viruses that are not cellular (they are protein complexes), but then they may not be living organisms.

To summarize, there are four criteria applied to any statement before it can be accepted as a scientific law: (1) The statement must be about kinds of events and not directly about any singular event. (2) The statement must show a functional relation between two or more kinds of events ("kind of event" refers either to things or to properties of things). (3) There must be a large amount of data confirming the law and little or none disconfirming it. (4) The relation should be applicable to very different events (although there may be limiting conditions).

Scientific Explanations

We have seen examples of data and of laws, or principles. We shall now consider some examples of explanations. Developing and elaborating explanations is a major part of the scientific enterprise. The problems of defining data and laws are difficult (and we did not resolve them), but the problem of explanation is even more so.

Scientists try to take the data in a given area and invent a general principle or set of principles with which these data are compatible. In other words, they attempt to develop a framework within which they can view events and data and understand them. In general, the greater the number and diversity of events that can be seen to be compatible with a relatively simple framework, the better the explanation. An important part of the game of science is to develop the simplest framework, from which one can generate the smallest set of hypotheses or principles that will account for the greatest variety of events. Once scientists arrive at seemingly workable principles, they then reverse the process and use these principles to indicate new events to be observed so that they can find out whether these new events are consistent with the proposed framework.

Consider astronomy. People have observed the heavenly bodies since antiquity. Any people who casually observe the skies and note what they observe can determine certain events and laws. On one night the observers will notice that the stars move gradually toward the west as the night progresses. On the next night they will notice again that the stars move gradually toward the west. If they continually watch the heavens, they become aware that the same patterns of stars appear regularly but that they do not appear at the same place at the same time. Each night a given pattern is at the same place in the sky about four minutes earlier than the previous night. Almost everyone who looks at the skies assumes that the same pattern means that the same stars are appearing again and again. Our observers notice peculiar stars, however, that appear near the same place among other stars on successive days but in the course of a longer period of time change their positions radically. These stars have been called "wanderers," or "planets." All of these observations and interpretations were made by the Egyptians and Babylonians, and they serve as a background set of beliefs and information necessary to our understanding of later developments.

As an example of an early attempt at *explanation,* consider Aristotle's discussion of these events. He suggested that the earth is a stationary sphere in the center of the universe. All the heavenly bodies are on huge, clear spheres that go around the earth in perfect circles. The sun, moon, and each planet have their own sphere spinning around the earth, and all the fixed stars are on a single sphere farther away from earth than the others. The different motions of the planets were explained by the assumption that different spheres travel around the earth at different velocities. All the heavenly bodies are perfect and unchanging, and each of them travels along one or more perfect circular paths.

This is an explanation of the facts as they were known to Aristotle. From it, one can ask certain questions and get answers. (1) Why do the same patterns of relations generally exist among the stars? Because most of the stars are spots on the same sphere. (2) Why is it that five stars, the sun, and the moon do not remain in the same place among the other stars? Because each of these seven heavenly bodies has its own circular orbit and rate of revolution around the earth, and these rates are slightly slower than the stars' revolution. The sun makes one complete revolution around the earth a day. The stars also make approximately one revolution a day. However, since on a given night they are about four minutes more westerly in the sky than on the previous night, they make one more revolution a year than does the sun. The moon makes about 13 revolutions fewer than the sun; and each planet goes around on its own schedule. From this theory, by knowing where a heavenly body is at any given time observers can predict where it will be at another time and "postdict" where it was at a previous time.

Aristotle's theory was not very powerful, because he did not have a simple framework that could be described by a small set of principles. Each orbit had to be independently plotted, and there was no general explanation of why the planets orbit at their particular rates. Furthermore, there was no indication of the nature of the huge sphere that houses the stars and is the ceiling of the sky.

Using this theory or similar ones as a basis, however, astronomers tried to plot more exactly the paths of the planets to establish how fast they travel in their circular routes around the earth. These astronomers found problems developing out of these explanations. Sometimes planets seemed to be traveling faster than the fixed stars, although generally they traveled more slowly. Two of the planets, Venus and Mercury, never got very far from the sun. When they rose in the east before dawn, they were known as morning stars; when they remained in the western sky after the sun went down, they were known as evening stars. From these observations of the planets, the astronomers concluded that none of them could be circling the earth at a constant speed. Simple circular explanations were not consistent with the observed facts. When an accepted explanation no longer explains the major facts well, scientists recheck them, and some usually try to modify the theory to handle the facts. Aristotle modified the theory by adding rotations within rotations. In this way he could still explain the heavens in terms of constantly moving spheres. Another possible modification could have been to assume that the planets did not travel at a constant speed. Sometimes they would move faster than the stars and sun, sometimes at the same speed, and sometimes slower. However, simply stating that the planets change speed is not a satisfactory explanation, since one needs an independent reason why the planets would speed up or slow down. When an explanation becomes solely a restatement of the data, it is not a scientific explanation. We are simply giving the phenomenon a name. Our example leads only to questions and answers such as "Why is Venus rising earlier and earlier daily? Because it is moving faster." "Why is Venus now rising later and later? Because it is now moving slower." These answers can be reduced to "It is moving faster because it is moving faster, and it is moving slower because it is moving slower," which is no explanation at all. An explanation has to be a conceptual schema that organizes and extends the data; its truth or falsity has to be testable. In other words, if all *possible* events automatically confirm a theory, the theory is worthless. Some possible events should tend to negate a theory.

A theory that can "explain" everything is very seductive. It can seem very attractive because, if you have in mind some data or expected data, you can always see how the theory could indeed "explain" it. Scientists have to be careful not to fall into the trap of devising such theories. Sometimes they might have what seems to be a reasonable theory, one that can account for some data but not others. In modifying the theory to account for the new data they might make the theory able to account for *any* possible data. If it reaches the point at which one cannot imagine any event or set of events that would imply that the theory is false, the story automatically becomes scientifically useless. This doctrine, called *the principle of falsifiability*, is extremely important.

Several well-known theories gradually added qualifications until they were no longer falsifiable. The doctrine that ether in space serves as a medium for light transmission is one; the theory of phrenology, according to which personalities can be explained by bumps on the head, is another; astrological prediction is a third; certain variants of Freudian psychology constitute a fourth. For example, we recall a psychoanalytic explanation of a child's misbehavior as almost always being the mother's fault. Sometimes the mother is around too much—she is overprotective. Sometimes she's not around enough—she is guilty of desertion and lack of love. Sometimes she's there at some times and not there at others—the child cannot handle the inconsistency. In this example, if the child misbehaves, the amount of time that the mother is with him or her will explain the misbehavior, regardless of what that amount is. Such a theory has no explanatory power at all.

The concept of falsification has received a good bit of attention both in the philosophy of science and in science. It is most closely identified with Sir Karl Popper, a very well known British philosopher of science. The distinction between scientific statements and nonscientific statements rests on whether there is some possibility of falsifying the statement. For example, take a statement by Duane Gish, a well-known creationist: "We do not know how the Creator created, what processes He used, *for He used processes which are not now operating anywhere in the natural universe* [italics in original]. . . . We cannot discover by scientific investigations anything about the creative processes used by the Creator." Not only does Gish dispose of the possibility of falsification, but of the possibility of any positive evidence as well. This is about as close to science as mud pie making is to the study of plate tectonics. Just how this leads to an explanation of Scientific Creationism (an oxymoron) leaves us more baffled than if it had been written in Sanskrit by an illiterate. Compare this with the bending of light in a gravitational field. Einstein predicted the sun would deflect light from stars by 1.75 seconds of arc. The Royal Society and the Royal Astronomical Society measured this during a total eclipse (May 29, 1919) and got 1.64 seconds of arc, well within the possible error of measurement. Einstein exposed his work to possible falsification and was triumphant.

If there is no possibility of finding data that tend to contradict a statement, you have no science. However, a problem arises in identifying the contradiction and determining its extent and importance.

After Newton, the idea of a sun-centered system was accepted by all but a few. Yet it was possible to raise an argument that seemed to falsify the whole notion. The problem was stellar parallax. The argument ran something like this: If you view a star from opposite sides of the earth's orbit it should appear at different distances from other stars; the pattern made by the stars should vary as the observer moves from place to place. Consider viewing a football game from different places in the stands. From the end zones the players seem to be horizontally spread out all across the field; on the 50-yard line most of the players might seem to form a vertical chain.

It was not until 1838, almost 300 years after Copernicus and 151 years after Newton's *Principia,* that Bessel found parallax among the stars. Parallax was difficult to measure because the stars are much farther away from us than almost anyone had suspected.

Einstein also had problems of apparent falsification on several occasions. For example, Walter Kaufman produced evidence that seemed to indicate that Max Abraham's theory was in better agreement with the data than was Einstein's. As was his usual practice, Einstein apparently paid little attention to the results. He didn't fault the work. Rather, he wrote a more general reply: "In my opinion other theories have a rather small probability because their fundamental assumptions concerning the mass of moving electrons are not explainable in terms of theoretical systems which embrace a greater complex of phenomena." Einstein soon was demonstrated to be right.

Where does all this leave the struggling, confused scientist? He has no absolute rules or answers in the back of the book. Darwin was perhaps the first to understand and offer a solution. In *Origin* he wrote:

> For I am well aware that scarcely a single point is discussed in this volume on which facts cannot be addressed, often apparently leading to conclusions opposite to those at which I have arrived. A fair result can be obtained only by fully stating and balancing the facts and arguments on both sides of each question. . . .

Today this approach is rather common among practicing scientists. Any finding or idea is embedded in a complex of assumptions, data, and ideas. There are no final, absolute conclusions. Rather, we look at evidence to see whether it forms a coherent picture. At any particular time it is a matter of judgment. If the evidence changes the judgment might change.

Let's return to explanations of astronomical phenomena. One that was accepted without question from the second century until modern times was that of Ptolemy. He modified Aristotle's position by claiming that the planets travel in little circles (epicycles) on their orbits while they are circling the earth. Explaining the paths of the planets this way enabled

Ptolemy to predict pretty well where the planets were likely to be at a given time. For accurate predictions, Ptolemy had to assume that the center of the orbit of the planets was not always the earth, but he arrived at no reasonable explanations for the eccentricity of the orbits.

During the 16th century Copernicus had a major insight. He studied and plotted the paths of the planets, but he could not fit the data (the planets' locations over time) to the Ptolemaic system. He attempted to explain the data by assuming that the sun, and not the earth, is the center of the universe and that the earth goes around the sun and spins on its axis, as do all the other "wanderers" except the moon. By assuming that the earth spins on its axis once a day, he could hypothesize that the fixed stars that seem to rotate were actually stationary. They would no longer have to be on a huge curtain that spins at a fantastic rate. Day and night could easily be accounted for by the earth's spinning on its axis. In fact, this theory had unexpected explanatory power; the year and the seasons could also be accounted for easily, and the motion of the planets could be accounted for, to a first approximation, by assuming that they also circle around the sun. Their seemingly complex path is due to the position of the earth when the planet is viewed.

Copernicus' explanation, which came after much hard work, directly contradicted both the religious dogma and the scientific explanations of the day. This explanation revolutionized human thought and was vital to the new learning of the Renaissance. The explanation, however, was not completely satisfactory.

About 50 years after Copernicus died, Johannes Kepler began trying to find a simple way to describe the paths of the planets. He spent years studying the data of Tycho Brahe and others and personally plotting the path of Mars. He assumed that the sun is responsible for the planets' paths, and that they revolve around it. But a circular path did not fit the data. Mars is not always the same distance from the sun, and putting the center of the circle elsewhere made no sense to Kepler. He tried many different ovals with the sun at the focus, and still he could not fit the path of Mars. Sometimes the planet's orbit was inside and sometimes outside of the ovals he tried. He finally tried to approximate the orbit with an ellipse. To his surprise, he could now predict the path without error (within the accuracy of his measures) assuming the sun at one focus. This prediction, after much more contemplation, led to Kepler's first law of planetary motion: planets travel in elliptical orbits with the sun at one focus. The data collected about other planets were consistent with this law. Given reliable paths for the planets, Kepler could now formulate a law to account for their relative speeds in their orbits. This law is: a line connecting the sun with a planet sweeps equal areas in equal times. The planet does not travel at a constant speed but travels faster when nearer the sun than when farther away. His third law—that the ratio of the cubes of the mean distances of any two planets from the sun equals the ratio of the square of their periodic (orbital) times—delighted Kepler and showed the power of the explanation, because it accounted for more than did any previous theory. It gave a systematic relationship to the speeds of the different planets, and it explained why some planets travel around the sun faster than others. From these three principles, with only a limited amount of data, one could generate the paths of all the planets and consequently explain the observed movement.

In addition to being descriptive *laws,* Kepler's laws led to *explanations* of the motions of the planets, because he related the motion of a planet both to the motion of other planets and to its specific relation to the sun. He had a conceptual framework in which all the planets are of a similar nature and their movements are consistently related to one another in quantitative precision. If one planet travels along a certain path, all planets will travel along similar paths, and so on. Thus he had a unified framework on which he could rest detailed explanations. In answer to the question "Why does Mars travel in an elliptical orbit?" he could answer that all planets travel in elliptical orbits. If we asked "Why does it travel faster sometimes and slower at other times?" he could answer that planets increase their speed as they near the sun; also the radius vector (a line extending from the sun to a planet) sweeps equal areas in equal times. To the question "Why does Mars travel faster than Jupiter?" he could answer that there is a general principle that the closer the planet's orbit is to the sun, the faster it travels, and this principle can be formalized with a specific function relating the planets.

Explanations are never complete. Kepler did not explain why planets travel in elliptical orbits except to say that it is part of their nature; nor did he explain why their radius vector sweeps equal areas in equal times. But Kepler did offer legitimate explanations. The principle "Mars travels in an elliptical orbit around the sun" is part of an explanation of why Mars is seen where it is in the sky on a given day. The principle "Planets travel in elliptical orbits" is part of an explanation of why "Mars travels in an elliptical orbit." The principle "The closer the planet to the sun, the faster it travels" is part of an explanation of why Mars travels faster than Jupiter and why Mars travels in its orbit at variable speeds. Kepler thought that the sun causes these behaviors, since it is the focus of the radius vectors. It can be seen that laws and principles are used to explain lower-order laws, and these laws themselves then become candidates for explanation. Kepler's laws were explained by Newton.

Newton, who was born 12 years after Kepler died, explained Kepler's laws by gravitational principles. Whereas Kepler had three laws of planetary motion that explained the movement of all the planets, including earth, Newton had three laws of motion that related such extraneous things as falling apples and the tides to the same principle as planetary motion. From Newton's principles, not only could Kepler's laws be derived, but most deviations from the elliptical orbits (too small for Kepler to see with his instruments) were also accounted for. Newton had a potential means to account for the discrepancies from his principles. It was thought that they could be accounted for by unknown planets. Astronomers looked for these. Due to perturbations in the orbit of Uranus, astronomers looked for another planet and discovered Neptune in 1846. A planet "Vulcan" was proposed to account for the discrepancies in the orbit of Mercury. This search came up empty. It took Einstein's theory of relativity to account for these.

Scientific explanation, which is at the core of the game of science, requires relating the laws describing a class of events to a framework and some set of principles that are described using terms other than those used for the laws. When the laws are seen to be derived from the principles, then scientific explanation is achieved. It must be noted that scientific explanation is more than a description of events, and it is more than a law describing the particular class of events. Explanation requires an interpretation of what the events are and an indication of the reason they occur the way they do by relating them to a unified framework. Scientists investigate carefully to see whether the events occur the way the explanation suggests they should, and they investigate other events that should occur the same way. Generalization to a different set of events cannot be done if the scientists only describe and do not give an explanation. In astronomy, a description is about spots of light that appear in the sky nightly, whereas an explanation is about solid bodies that travel around another body in elliptical orbits because of the gravitational field surrounding all bodies.

Obviously, as a science advances, new explanations develop in an attempt to explain the same kinds of events. But there is an important consideration in this advance. Although astronomical events were essentially the same in Aristotle's time as in Newton's, the data were not. As the technology of observation of the stars improved and more people systematically plotted their movement, the data changed considerably. A good scientific explanation is one that is in accord with a great amount of data. The data abstracted from the events are the test of an explanation; the events themselves are not such a test.

We can see from the preceding lengthy example that people have been fascinated by the heavenly bodies for over 5000 years. Why this interest? They had a puzzle. The puzzle was laid out above them every night. It naturally piqued their curiosity and challenged their ability to solve it. Were there any other reasons? Probably. Does it matter?

The Public Nature of Science

We have seen that science depends on individuals' observing facts, collecting data, and coming to their own conclusions about how to organize the data. However, it is also important that scientists communicate their findings and explanations to others. Although a single individual living on a desert island could do science, science is not a game of solitaire; it has many participants. An explanation does not aid the science until other scientists can be convinced that it is a useful explanation.

The reasons for the necessity of convincing others of a position are threefold. First, by attempting to present data and argue for an interpretation, scientists have to spell out clearly what the data are, what the explanation is, and what the reasons are for believing the explanation. The scientists themselves may gain new insights from this intellectual activity. In spelling out a position, they often find unexpected problems in the explanation, and they may thus find that further work is necessary to clarify the position. In order to present a position to others, scientists must evaluate their position objectively. Insights accompanied by feelings of exhilaration are not enough. They are not always associated with valid explanations. During an objective evaluation and formulation scientists may convince themselves that an explanation that they thought valid is not. Of course, evaluation does not always lead to disenchantment with the tentative explanation. The scientists may, to their delight, find that they can extend the explanation into unexpected areas, or that slight modification will enable them to do so.

Second, the output of a scientist must be put on public display before it can be considered a scientific contribution. This requirement makes scientists careful in their formal theorizing, since an ill-conceived or poorly supported hypothesis can be (and often has been) a lifelong embarrassment when it endures in cold, unerasable print. By the nature of the game, scientists' work has to be open to public scrutiny; the only method of gaining acceptance requires having their work approved by their scientific peers. If scientists present their data and explanations but cannot convince some peers of their validity, the scientific community as a whole will tend to reject their claims. The scientific community requires an explanation of data to be acceptable to a sizable minority before the privileges of scientific consideration are extended, such as publication in scientific journals and inclusion in textbooks and monographs. Without this exposure, a theoretical position has little chance of directly contributing to the science itself.

The third major reason for communicating scientific discoveries is simply to inform others about them. Convincing oneself and others of the validity of the discovery may be important in the advancement of one's own knowledge, but society at large should also be informed. Only after scientific knowledge is communicated to applied scientists and practitioners can it be used to benefit society. If the principles of electromagnetic waves had not been known to Guglielmo Marconi, he would have had great difficulty in developing the radio. If the principle of universal gravitation had not been known, it could not have been used to develop artificial satellites. Also, without free and open communication different people would spend years working on the same problem. There is a great waste of both effort and scientific ability if a scientist has to go through the same steps that someone else has already been through. Finally, only through free communication in books and the mass media can the general public learn about themselves and the world. In the modern world, the general public indirectly supports scientific research. This public should know something about what it is buying.

Scientific Methodology

We have seen that science has as its object of study the natural world and that scientists strive to explain the facts of nature. We have discussed briefly what facts and data are and what constitutes an explanation. Our next problem is to examine how the scientist goes about finding facts and establishing explanations of them.

When scientists are at work they are not likely to gather data by casually observing the world around them. They may start by casual observation, but data obtained in this manner rarely show the relationships necessary for a powerful explanation. Such an explanation requires precise and reliable observations. Casual observation is likely to be contaminated by many extraneous factors.

Obviously, the explanation of planetary motion established by Kepler could not have been done without carefully controlled observation. The information he used had to be detailed and accurate. In addition to all his hard work, he had very good luck. In planetary motion, the effects of all variables except those caused by the sun are quite small. The elliptical orbit, which describes the path that a planet would take if the sun were the only influence on planetary motion, is the dominant orbit. The gravitational fields of the planet itself and those of other planets are quite weak compared with that of the sun. By using the observations of Tycho Brahe and observing as carefully and objectively as he could, Kepler plotted the orbits of the planets. Kepler identified the planets' compass direction from his observatory, and their height above the horizon for different times, days, and years. Using the insight he attained from Copernicus, he worked at converting these measures to orbits around the sun. He needed to develop a system that would be consistent with all careful observations of planetary motion, observations made by others as well as himself. It is obvious that if he did not record the time of the measure accurately, or the date, or the compass direction, or the height, or the location on earth from which he measured, he would have anomalous data—that is, data that would not fit a theoretical orbit of a planet. Careful and accurate measurements were essential. Thus, with hard work and insight, Kepler formulated his major laws of planetary motion. The naturally occurring orbital paths are almost identical to Kepler's theoretical ones.

Aristotle, another great scientist, established many of his scientific laws on the basis of casual observation. One of his great laws was that heavy things fall faster than light ones. This generalization, which had much observation to support it, was accepted as a fact for almost 2000 years. Anyone knows that rocks are heavier than feathers and that they fall faster. In fact, feathers don't even fall; they float. One problem with Aristotle's law is that falling bodies are affected by many things. In a very famous demonstration (probably apocryphal), Galileo went to the top of the Leaning Tower of Pisa, and before a large crowd simultaneously let a heavy ball and a light ball fall. To everyone's (except Galileo's) surprise, the balls stayed next to each other during the fall and landed together. By setting up an artificial situation, Galileo provided data that allowed him to establish an important natural law.

A paradox within science is that artificial, or controlled, situations help scientists understand the natural world. In natural settings many aspects of any two events differ; the scientist sets up artificial situations to make events similar to each other in many ways. Consider the example of falling bodies. Many events influence their fall—the impetus with which they start falling, the things that divert their fall, their shape, the amount of air they displace, as well as (possibly) their weight. Also, the difference in their rates of fall may be slight. If two things fall at different times from different heights, observers may not know which actually fell faster. Even if they do, they may not know whether the height from which they started to fall made any difference. The story is that Galileo contrived the time, place, and condition of the fall; only by doing so could he discover certain natural laws. By dropping two balls of different weight from a great height at the same time, he controlled the height of the fall, the time of the fall, the shape of the objects, and the material of which they were made. By letting the objects fall a long distance at the same time, he could determine even slight differences of speed. When they fell together, he could easily conclude that weight per se does not influence the speed of falling objects. Observations in natural, or uncontrolled, settings had never led to this conclusion.

To study this phenomenon Galileo actually built long, smooth inclined planes and rolled balls down them. He could have watched logs and rocks roll down hills, but again many factors would have been different at different times. Galileo worked for years carefully releasing balls and timing how long it took them to cover different distances. By controlling the situations—that is, making them artificial—he could determine and study the important relations. Like Kepler, he followed false leads and had trouble establishing simple relationships, but he finally came to important conclusions concerning motion. One conclusion was that a constant force does not result in a constant velocity; rather, it results in a constant acceleration. Time and acceleration were seen to be more important variables than velocity and distance. These important discoveries could not have been made if Galileo had not invented artificial situations in order to control and accurately measure the effects of important variables.

Since there are so many things happening in the real world, explaining why and how things happen requires creation of situations in which less is happening. Once laws have been established in simple situations, combining the effects of different variables is usually less difficult. However, even this part of the game may not be won. For example, we know how to describe the path of an object coming near a stationary object, but we still cannot solve the laws of the paths of three moving objects passing near one another.

The use of relatively simple, controlled, artificial situations to study the influence of certain variables in well-defined contexts is particularly necessary in the biological and the social sciences, because the influences of extraneous variables may not be so obvious as in the physical sciences.

An example of the importance of control and artificial situations in biological sciences is the formulation of the genetic theory of heredity by Gregor Mendel and its continuing extension by Barbara McClintock and others. It had long been known that certain traits of one generation are passed to another. However, the explanation of this phenomenon was rather vague and was somehow attributed to the blood of the parents. This is where such expressions as *full-blooded* and *half-blooded* come from. Even plants supposedly transmitted traits to their offspring through a fluid in some manner. Both plant and animal hybrids had been grown and bred without the processes involved being understood. Certain individuals had spent many years investigating hybrids and had some knowledge about them—for example, they knew that hybrids tend to return to pure species over generations—but none had a reasonable theory to cover such phenomena.

In 1857 Mendel began to study different varieties of garden peas. He bred them by carefully controlling their pollination for about eight years before reporting his results. His paper furnishes us with an example of careful experimental control and can still be read with profit. (It is reprinted in Newman's *World of Mathematics*.) In each experiment, Mendel considered for study only two varieties of a single species. In one experiment, these varieties greatly differed in size, one being 6–7 feet tall and the other only ¾ to 1½ feet tall. He inbred these strains for two generations to confirm that they were stable varieties. This was done by demonstrating that all offspring emulated their parents. He then made a hybrid strain by dusting the pollen from the tall plants onto the eggs of the dwarf plants and vice versa. He covered the plants prior to their dusting so that no other pollination could take place. The resulting hybrid pea plants were all at least as tall as the tall parent plants.

Mendel called the trait that was transmitted to the offspring *dominant*, and the trait that disappeared *recessive*. He used the term *recessive* because the trait that disappeared in the first hybrid generation reappeared in successive generations. For the next generation Mendel self-pollinated all of the plants. That is, he dusted the pollen of a plant onto the plant's own eggs, and then planted the resultant seeds. The new generation included both tall and dwarf pea plants. About ¼ of the plants had retained their grandparent's dwarf stature. The other ¾ of the plants were tall like their parents.

Description of the results of self-fertilization of the plants from this generation is somewhat more complex. The offspring dwarf plants were all dwarf. About ⅓ of the tall plants had offspring that were tall. The other ⅔ of the tall plants had both tall and dwarf offspring in the same mix as their parents did, about ¼ dwarf and ¾ tall.

In later generations self-fertilized dwarf plants had only dwarf offspring. Self-fertilized tall plants with only tall siblings had only tall offspring. About ⅓ of the tall plants with both tall and dwarf siblings had only tall offspring, and the rest of these had both tall and dwarf offspring in a ratio of about 3 to 1, respectively.

Later experiments combined plants that differed in more than one factor—for example, the form of the fully developed seed (round or wrinkled), and the color of the seed coat (yellow or green). Mendel concluded that the different traits in hybrid union, as represented in their offspring, were totally independent of one another. He reported that all possible combinations occurred with the mathematical probability identified by random combinations of dominant and recessive factors. These latter results have been modified by later research. We now believe that, on occasion, different traits are linked to one another.

Mendel proposed that both the pollen and egg cells contain hereditary factors (called *genes* since about 1909). Each characteristic of an organism is represented in both its pollen and its eggs by one of these hereditary factors. After fertilization, the seed contains the factors from both sources. In constant strains, the genes from the pollen and the egg are the same. In hybrids, corresponding genes represent different traits. Thus, when a tall pea plant and a dwarf one were cross-fertilized the resultant seed contained both factors. In peas, the tall trait is dominant, so the dwarf factor, although present in the seed, could not become manifest in the plant grown from it. This recessive factor, however, has the same chance as the dominant one to be represented in the pollen or egg cells of that plant. In self-pollination, this hybrid randomly combines some pollen containing the tall factor so that some of it combines with eggs containing the tall factor, and some of it combines with eggs containing the dwarf factor. Pollen containing the dwarf factor are also randomly combined with eggs containing each factor. Some of the resultant seeds will contain only the dwarf factor, some will contain only the tall factor, and some (approximately twice as many) will contain both factors. Other factors combine in the same way, and according to Mendel, all such traits were independent of one another.

With well-designed experiments such as the ones Mendel performed, one is in the position to find out just how different aspects of a system affect one another. This kind of specific information aids the development of theories that account for the phenomena and relations studied. It would have been difficult, if not impossible, for Mendel to invent such an insightful theory without having a consistent and constrained data base on which to theorize.

It is important to note, before we leave this section on scientific methodology, that although experimental methods are extremely important, and the most efficient means of establishing functional relationships, they are not the sole means of gaining important scientific information. Many sciences use careful observation as a means of obtaining critical data, from which they may invent and support major theoretical insights.

A critical factor involved in scientific methodology is to define a relatively constrained domain within which to evaluate a meaningful or important question or set of questions. Questions are meaningful to the extent that they are central to the structure or function of the phenomena under study. Scientific research is difficult, in part, because the domain should be a coherent "system" in nature. That is, the things studied should have relatively strong internal relations among one another, and the whole system should often act as a unit among other factors not under study. Sometimes people disagree because it is not obvious where these "joints" in nature are. Major advances often occur when a scientist identifies a set of relationships in one observational context, that can be shown to generalize across many contexts. Mendel, for example, studied the height of pea plants and the color of peas, but he identified some relationships that hold from bacteriophages and fungi through maize, mice, and human beings.

Once the domain is identified, it is important to find an observational context and to observe all of the relevant information within that context. Then the scientist attempts to build or extend a coherent framework that incorporates the collected data as a natural consequence.

Mendel successfully created his theory by identifying a domain of research (successive generations of a single species), carefully constraining the organisms he investigated, and controlling the genetic factors that affected them. His work took place over several years in a corner of his monastery. Darwin took the variability of species as his domain. The question he considered was "What is the origin of the numerous species?" He carefully observed the similarities and differences of species and the environmental niche in which they were successful. He did not control or manipulate his data, but he carefully studied it and compared relations among species from all over the world.

Kepler took as his domain the motion of the planets. His main observational context was the orbit of Mars. His goal was to find a simple regular description that could account for the location of the planets in the heavens at all times. He also observed and compared, but he did not manipulate the phenomena he studied.

We have seen that scientific advancement often depends on artificiality, selection, and control. The scientist, in order to investigate the effects of certain variables, sets up conditions under which other critical variables are not allowed to have an unpredictable effect on the experimental results. By observing the behavior of objects under artificial but well-specified conditions, the scientist can formulate laws that express this behavior. Artificiality and selection, rather than detracting from understanding the events in the real world, create conditions by which one can understand that world. The real world has so many things happening at once that even geniuses such as Galileo or Mendel had to create artificial situations in order to understand natural processes.

Selection and control may be the most powerful methods for attaining knowledge about natural phenomena, but there are many instances in which selection and control seem impossible. Scientists in such areas as archaeology, ethnology, paleontology, political history, sociology, and astronomy cannot manipulate the situation to suit their own purposes. Their only control is to select what and how they observe. These scientists strive to understand phenomena. They invent general principles and attempt to develop a framework within which their data are compatible. For confirmation they try to extend their predictions to include other data. They are limited in that the phenomena are always complex, and they cannot always observe when and what they want. Their task in many ways is much more difficult than that of the experimental scientist, because different events in nature vary in more than one way.

Many sciences present problems that cannot be put into a formal experiment for one reason or another. With some of these problems, it is possible to use statistical techniques to isolate the effects of some of the variables. For example, statistical information confirmed the theory that cigarette smoking leads to lung cancer and certain heart conditions.

No matter what methods scientists use to isolate the effects of variables, their aim is always to achieve a broader understanding of natural processes. And it is this understanding that is the primary goal of science.

The Uncertainty of Science

Although scientists seek understanding, they never achieve it completely. The game of science never ends; all conclusions are tentative. No matter how much information scientists have, they can never be certain of any of their conclusions. There are vast areas in any science where even the working scientists are quite uncertain about how or why events happen the way they do. Most scientists are relatively certain that some currently accepted laws and explanations are quite accurate, but they have no guarantees. The problem is at base a logical one. Philosophers of science currently believe that certainty in the natural sciences is a logical impossibility. They may allow that some areas of endeavor are certain within logic and mathematics, but never in the real world. Even in logic and mathematics, however, there are statements that cannot be evaluated as well as others.

The 18th-century philosopher Immanuel Kant concluded that some things about the world are known for certain—for example, that the sum of the angles of a triangle equals 180° and that all change is continuous. But let's examine these "certainties." If a triangle is made of wood and the angles are measured as carefully as possible, the angles will not consistently total 180°. Moreover, if a large triangle is conceived as having two stars and the earth as its vertexes, and if straight lines of the sides are measured by light rays (light supposedly travels in a straight line), the angles do not always total the same. In fact, the totals vary significantly. We can conclude from these facts either that light does not always travel in a straight line or that the three angles of a triangle do not always add up to 180°. In neither case are we certain that any physical entity duplicates a Euclidean triangle. We do know that the three angles of a Euclidean triangle add up to 180°, but in practice we can never be certain that we have constructed a Euclidean triangle.

Kant also "knew" for certain that all change is continuous. He meant by this that there cannot be any really sudden changes. You know, for example, that, if you are traveling in a car and suddenly slam on the brakes, the car does not stop immediately; it takes about 210 feet to stop a car traveling at 60 miles per hour. If a rock falls, it gradually increases its speed. Kant assumed that by reason alone he knew that *all* change is similarly continuous. Even an explosion is supposedly a gradual, though rapid, expansion of gases. We are still not certain whether all change is continuous, but we *are* certain that we don't know by reason alone that all change is continuous. The question can be debated. In opposition to Kant, current physical theory states that in small systems, such as atoms and molecules, changes are in discrete units. According to this theory there is no such thing as one and a half molecules of water or five and a half photons of light. In addition, if someone turns on a flashlight, the light *immediately* travels at full speed rather than gradually increasing to maximum speed. Even electrons in their orbits around the nucleus of an atom seem to go from one orbit immediately to another, spending no time getting there. It seems just as easy to believe that all change is discrete as to believe that all change is continuous; we cannot know without experiment which is true. And, since no experiment ever provides absolute certainty (because there are always *some* uncontrollable variables), we cannot know *for certain* which is true.

We do have some certain knowledge. For example, the summation 2+2=4 is derivable from definitions. However, we don't know for certain that 2 apples put in a basket with 2 other apples will yield 4 apples. We are pretty sure they will, but we're not absolutely certain.

This is a very difficult concept to grasp, but it is important in understanding the limits of science. There are two ways we can determine with certainty that a statement is true: (1) we can define it as true, or (2) we can derive it logically from statements that are defined to be true. Since we can't derive any true statements about classes of events in the real world from definitions alone, no scientific statement is true with certainty. For example, we can define the concept *material object* as something that does not suddenly appear or disappear, and we can define the thing in front of us to be an apple. But we cannot then *define* the apple to be a material object; we have to *investigate* to find out whether it is one. And since we have to investigate to find out, we can only know *inductively* that the apple is a material object, and nothing that is known only by induction is known for certain, as is explained and illustrated below.

If we analyze any law or explanation in natural science, we find that at some level it ultimately rests on *induction*. In other words, at some level an assumption is made that, since an event has occurred before on several occasions, under similar conditions it will happen again. One reason that no conclusions are certain in science is that there is no way of knowing *for certain* that the same thing will recur the same way. A low-level example of induction is that we expect the sun to rise every day. We may not see it because of clouds, but we expect it to be there. Why? (1) We have been told that it will rise on a regular schedule, and we may even have had the theory behind that schedule explained to us. (2) We know that it has risen every day of our lives, so why should it stop rising?

Unfortunately, these two reasons are not enough to tell us for certain that the sun will rise tomorrow. (1) Other things we have been told, and have believed, have not always turned out to be true; furthermore, the theories accounting for the things we have been told may be wrong. (2) The evidence that an event has occurred with great regularity is not certain proof that it will continue to do so, as we can see from a short parable. Once upon a time there was a very intelligent turkey. It lived in a pen and was attended by a kind and thoughtful master. All of its desires were taken care of, and it had nothing to do but think of the wonders and regularities of the world. It noticed some major regularities—for example, that mornings always began with the sky getting light, followed by the clop, clop, clop of its master's friendly footsteps. These in turn were always followed by the appearance of delicious food and water within its pen. Other things varied: sometimes it rained and sometimes it snowed; sometimes it was warm and sometimes cold; but amid the variability, footsteps were always followed by food. This sequence was so consistent it became the basis of its philosophy concerning the goodness of the world. One day, after more than 100 confirmations of the turkey's theories, it listened for the clop, clop, clop, heard it, and had its head chopped off. Thus, regularity does not guarantee certainty, and all induction is based on regularity.

Certainty is difficult to attain; not only can we not be certain of the same situation leading to the same results, but in actuality the "same situation" occurs only once. Assume that we are testing the boiling point of water. We pour some distilled water into a beaker, put a thermometer in the beaker, place the beaker in a stand, light the Bunsen burner under it, and wait for the water to boil. After it boils we can read the thermometer. Now assume that we wish to do the same thing the next day. Do we use the same water? It is now a day older. The same beaker? It has been used one more time to boil water. The thermometer also is older and has been used before. If we use different water, a different beaker, and a different thermometer, we do not have the same situation anymore. Furthermore, it is a day later, the sun is up for either a longer or a shorter period of time, we are a day older, the earth has rotated one more time, people have been born and have died, and so on. You may feel that none of this makes a difference; but that is an *assumption you make,* as does the scientist. The consistency of the data is one way to confirm whether the assumption is essentially correct.

Scientists try to duplicate what they deem to be the significant aspects of a given situation, but each situation is unique, and they do not even attempt to duplicate the total situation. Even the relevant attributes of any situation cannot be duplicated exactly. Scientists accept a certain range of values as duplication. If they are trying for quantitative laws, they will try to limit the range of variation as much as possible, but it cannot be eliminated completely.

We see continued improvement in accuracy of measurement. In all of science, probably the one measure that stands as the most absolute is the speed of light. In late 1972, the accuracy of measurement of that speed was increased one hundredfold by Kenneth Evenson and his research team at the National Bureau of Standards. That research team estimated the speed of light to an accuracy of 0.5 meter per second. In 1983 there was an interesting switch on this topic. Previously, the length of a meter was determined independently, and the speed of light specified by meter length. Now, because the speed of light is considered to be a stable constant, it is used to define the length of a meter. A meter is defined as the distance light travels in 1/299,792,458ths of a second. Any straightedge of that length is one meter long. Obviously we need indirect and inexact methods to make a measuring stick one meter long.

Another constant in an "exact" science is Avogadro's number. This is the number of molecules in one mole (gram-molecular weight) of a substance. The number is given as 6.02486×10^{23} molecules, but there is a 0.0027% error in this estimate. That is a very small error, but it comes to about 16,000,000,000,000,000,000 molecules. That is, Avogadro's number is accurate to about 16 quintillion molecules.

This lack of complete accuracy and certainty is no cause for alarm. It simply means that many of our answers can be improved. There are theoretical limits to the accuracy of our measures, but we have not reached those limits. Besides the limits of accuracy in measurement, the scientific explanations that we currently entertain are not certain to remain intact. However, it is unlikely that many scientific laws will change significantly in the near future, and whatever changes are made will come from individuals who are committed to one of the arenas where the game of science is played. The professionals usually score the points. Successes in the game may eventually be superseded by other successes, but in spite of that possibility they are exhilarating when they occur.

In this chapter we have examined some of the rules and concepts of the game of science. Basically we found that scientists have a systematic method that they use to get information and that they use their understanding of scientific methods to evaluate the work of others. Scientific knowledge is gained solely by the use of scientific methods. These methods are never certain, but they are the best we have.

5 Ideas and Their Development

Where do ideas come from? They may spring forth full blown and ready for battle like Athena from the head of Zeus, but more often they appear as weak, tottering, confused babes requiring time and nourishment before they can survive independently of their parents in a fiercely competitive world.

This discussion of the origin of ideas is highly speculative, although we can consider descriptions of the origins of ideas as data. However, what is needed is not merely an accumulation of data but some way of organizing the data into a meaningful pattern. Historians, philosophers, and psychologists have analyzed creative thinking and the formation of concepts and have tried to give them a meaningful conceptual schema, but no present conceptualization can account for a significantly large percentage of the data. Despite this limitation, however, a brief consideration of the problem is worthwhile.

Our discussion is divided into three parts: first, a consideration of the kinds of people who make contributions to the game of science; second, a very brief history of science; and third, an attempt to organize at least some of the conditions surrounding the production of new ideas. Wherever possible, the discussion is restricted to new ideas in science, since even this limited area is terribly complex.

The Kinds of People Who Contribute

Today, when the amateur scientist is nearly extinct, the primary contributors of new ideas are those formally trained (or, more rarely, educated) in a particular discipline. We can infer at least two characteristics of innovators from this fact: (1) They have spent a very substantial period of time and effort absorbing the laws, legends, and lore of a particular discipline while achieving a graduate degree. (2) They tend to have a high degree of intelligence, at least as measured by ordinary testing methods.

Even though adequately trained and relatively intelligent, the ordinary science-trained Ph.D. is still unlikely to make any substantial new contribution. Estimates vary somewhat, but a popular estimate is that about 10% of the individuals in a field contribute more than half the scientific publications. About half the Ph.D.s publish at most a single article during their career; those without Ph.D.s are unlikely to publish at all. Although published articles don't usually contain new ideas, publication is almost always necessary for communicating one. Unpublished material, no matter how original, has almost no chance of influencing the direction of science.

A great many scientific innovations have been the work of young scientists. Karl Gauss made contributions in his teens; Agnes Pockels discovered surface tension at 19; Newton's important contributions started in his early 20s and Einstein's in his mid-20s; Galileo discovered the regularity of pendulum vibrations at 17 and at 22 published work on the center of gravity of solids; and Candace Pert discovered endorphins while she was still a graduate student. There are some obvious reasons why youth has been favored. One simple reason is that there are more young people than older ones. Some contributors died young; certainly some potential contributors did also. Another reason is that young people generally tend to be more daring than their elders, perhaps because the young haven't seen as many "good" ideas fail. A few painful experiences do encourage caution. A third reason is that the young have not spent so many years working in a system and thus do not have the same personal investment in the status quo.

Studies of some other primate groups indicate that the old males are most resistant to change and least likely to display curiosity. For example, Japanese monkeys were observed during the introduction of new foods and behaviors into the troop. Typically one or more young monkeys responded first to the new situation, and other young monkeys then followed. The elders, particularly the males, either were last to respond or did not participate at all. Studies on a variety of animals have shown that older animals explore less and display less curiosity than younger animals of the same species. There are some problems with the studies mentioned, but they are our best evidence, and studies of monkeys are in general agreement with studies of scientists.

Whatever the reasons, the young tend to be innovators more often than their elders, and contrary to the primate research, men more often than women. There seems little doubt, however, that if it weren't for cultural restrictions, women would contribute more to science, and perhaps older women would contribute more than older men.

Some personal characteristics appear to differentiate the more creative scientists from the less creative ones. Creative scientists have a wide range of interests, which may vary over the whole range of human experience. What is sometimes reported as narrowness is quite often a lack of sociability, a limited tolerance for trivia, or interests that don't fit the usual pattern. Creative scientists may tolerate authority, but they definitely do not encourage it. Consequently, they often have problems with authority figures. Creative scientists' poor rapport with administrators and other sources of authority seems to be related to characteristics variously described as ego strength, arrogance, or detachment. They seem to have a greater than average ability to tolerate ambiguity and confusion in their work, at least temporarily.

Scientific creativity is related to personality factors other than sheer intelligence. We can only speculate about these factors at present, although research on them is currently under way at the University of California and a number of other institutions.

Personal observation and various surveys of scientists agree on at least one important personality characteristic of creative scientists: they are independent. This characteristic may not extend to all parts of their lives, but in their own fields they operate as individuals. A new idea is typically the product of a single person. To arrive at the idea, one has to withstand pressures to think as others do. Barbara McClintock's ideas were so independent that she was considered by some to be crazy, and she didn't give public seminars at her own laboratory for years because she was received so badly. Most scientists are not quite that independent. Also, a team may be useful if the members stimulate one another by providing different views and information that each can attempt to incorporate into his or her own personal framework. Contact with other scientists for discussion is also useful in the development of an individual's ideas, since this contact forces one to formulate those ideas in some communicable form.

Since particular ideas are individually produced, clearly the creative scientist has to be independent. Pressures to conform in the realm of ideas are just as severe as they are in dress, speech, or other behaviors, and just as hard for most people to resist. Incidentally the true nonconformist is *not* one who thinks differently from the majority in order to be aligned with a sizable minority. Instead, nonconformists have unique viewpoints for their own unique reasons. Nudists, John Birchers, and Theosophists are all conformists in their own way.

Historical Illustrations

In order to achieve a historical perspective, we need to abandon our view of the individual temporarily. From a distance, science is seen not as a smoothly flowing discovery of what is "out there"; rather, there are false starts, periods of stagnation, and major controversies, as well as great leaps forward. Within the major developments, there are innumerable smaller movements, countermovements, currents, and crosscurrents. Fascinating as these smaller movements may be, we shall ignore them; any detailed history requires volumes, and we have only paragraphs.

To examine cultural influence, let's consider the scientific efforts of the Greeks and Romans. For two or three hundred years the Greeks lived in a period of great intellectual achievement. Science was not the major product of that period, but Greek science was surprisingly successful. Part of this success was due to major achievements in mathematics. As pointed out earlier, mathematics is not a natural science, yet it is necessary for natural sciences. It is an exact expression that describes the world inexactly. Mathematical acumen can be seen in the development of astronomy. Greek astronomers reached a point far beyond their predecessors, the Egyptians and Babylonians. Considering the crudity of their instruments, the Greeks' achievements were fantastic. For example, Eratosthenes knew that the earth was a globe and calculated the diameter about 240 B.C. and was correct to within a very small error. A celestial system very similar to that of Copernicus was hypothesized by Aristarchus and his successor Seleucus. Such hypotheses have not always been well nor widely appreciated. Plutarch reports that at least some Greeks felt Aristarchus should be indicted on a charge of impiety for supposing the earth to revolve and rotate. (Galileo, accused on similar charges almost 1900 years later, might well have remarked "There is nothing new under the sun.") Late in the Greek era Archimedes used methods that resemble very closely those of modern science. Unfortunately Greek inventiveness largely expired with the coming of the Romans.

The great ancient empire of the Romans is an enigma. If one had to specify the conditions that should lead to scientific advancement, he might suggest a good intellectual beginning, a stable and relatively wealthy economy, places where different points of view are represented, and a common language for a large group of people. The Romans had all of these, but there is no evidence of any real achievement in science. The Romans were great engineers, soldiers, organizers, and traders; however, Roman scientific efforts and achievements as well as creative efforts in other areas were almost nil. They produced many great administrators but no great intellects. (Is there a conflict between these two roles?)

From a scientific point of view, the period between the Roman collapse and the rise of modern times, with little exception, is too dreary for serious comment. The beginnings of science in its modern sense can be dated from any of a wide number of different events ranging over several hundred years, with about equal accuracy. We could use the year 1210, when Aristotle's *Physics* became sufficiently well known to be banned in Paris, or 1245, when a knowledge of Aristotle was required for a Master's degree in Paris. Some judge early contacts between Western Europe and Islam, beginning with the bloody invasion of 1097 and culminating in the quiet seduction of the Western barbarians by a combination of Greek and Arabic learning, to be the critical period for science. Leonardo da Vinci (1452–1519) can also be identified as one of the originators of science.

At any rate, by the middle of the 16th century, very important developments were taking place. In 1543 Vesalius and Copernicus published their major works in anatomy and astronomy, respectively. The 17th century produced a flood of new concepts and findings. Even though a number of historians look at the 18th century as scientifically dull, it seems so only when compared with the 17th and 19th centuries. Historians have generally regarded the most fruitful scientific areas in the 17th century as astronomy (Galileo and Kepler) and physics (Galileo and Newton); however, biology also made great strides (Harvey and Leeuwenhoek). By the 19th century, science was moving forward in nearly all areas known at that time.

Consideration of the history of science has brought us to a few tentative conclusions. First, there are very few examples of isolated contributions. Stimulation from others is important in the generation of ideas, and new concepts tend to be grouped in time, such as the 17th and 19th centuries, and within groups where communication is relatively easy, as in Western Europe or Greece. Second, there are some things about a particular culture and its way of thought that may either facilitate or retard scientific development. This influence has been termed the *zeitgeist*, or the spirit of the times. There does seem to be some such factor operating, but we don't know any of its defining characteristics or exact influences.

Production of New Ideas

The third part of our examination of the origin of new ideas requires somewhat more detail than do the first two. For purposes of organization, we can arbitrarily break down the development of new concepts into four stages: (1) A problem arises in relation to an accepted concept, and that problem must be evident in some way. (2) The scientist involved has considerable knowledge in the area involved. (3) The previously accepted concept is reformulated or a new concept substituted. (4) The new concept must be developed to healthy proportions. This representation of the development of a new idea is obviously oversimplified, and these points form only the barest of bones, but they can be given some flesh through individual consideration. The points can be illustrated by aspects of the formulation of the theory of evolution.

The first stage in the development of the concept of evolution is the problem of how to account for the presence of all the diverse species of living creatures. This problem was recognized at least as early as about 450 B.C. by Anaxagoras, Democritus, and Empedocles. They raised the question of the origin of species; some entertained vague notions of evolution, while others had theories of the special creation of species. The latter theories eventually became established in Western culture primarily on religious grounds and provided the accepted conceptual framework that eventually broke down. As in most cases, the breakdown began when evidence appeared that seriously strained the capacity of accepted concepts. Part of the evidence was the discovery of a fantastic number of new species together with an increasing knowledge of their anatomy. The ark slowly sank under the weight of a myriad of hooves, paws, claws, and assorted feet. The fact that skeletal structure, internal organs, sensory organs, and musculature of such diverse creatures as cats, dogs, and horses bore striking resemblances was puzzling; humans also obviously had a great deal in common with many animals. Indeed, it was difficult to find any aspect of human anatomy that did not bear close resemblance to some animal.

The geologists formulated the idea that changes in the earth's surface were orderly, not catastrophic, and took place over great periods of time. This was a major burden for special-creation theories and provided working room for an evolutionary process. It is much easier to envision an evolutionary process occurring over millions of years than over thousands. Ideas grew that strata of the earth's crust were deposited slowly, at different times, and could be dated by their fossil content. These notions not only stretched time but also cast doubt on the possibility that the number of species remained constant. Or could there have been repeated creations of species? These and other lines of evidence forced the special-creation theory into extreme contortions.

The difficulties in any accepted concept not only must be real but also must be recognized. Humans, including scientists, have a peculiar ability to blind themselves to disagreeable or uncomfortable facts or ideas. Many scientists failed to see the limitations of the special-creation theory of species as late as 1900. In particular, Germany's Rudolf Virchow and France's George Cuvier, each of whom was a highly regarded scientist in his own right, remained unswayed by the evidence against the special-creation theory throughout their lives. In the United States the stronghold of scientific resistance to a change in the notion of the special creation of species centered around Louis Agassiz at Harvard. It is not completely clear whether Harvard of that day objected more to an attributed kinship to other primates or to other humans. On the balance, however, enough scientists saw great difficulties in the theory of special creation to arouse interest in and later acceptance of the theory of evolution.

The second stage in developing any new concept involves the background of the participating scientist. As previously stated, a scientist needs substantial information in a particular problem in order to formulate a meaningful new concept. This information does not necessarily lead to a new concept, of course. Science has its full share of walking encyclopedias who have a detailed knowledge about a particular area and yet leave the area totally undisturbed by new ideas. Only a few knowledgeable people develop useful new ideas.

Charles Darwin had attained a great deal of knowledge before he formed the framework for his evolutionary hypothesis. He was familiar with a considerable range of evidence from different sciences such as biology, geology, and paleontology. In short, Darwin, like other innovators, stood on the shoulders of his predecessors and contemporaries. They formed an important part of his grounding in the range of evidence surrounding the problems of species. In addition, Darwin himself was an excellent and careful observer and had the opportunity to make many new observations during his five-year voyage on the *Beagle.* Although formal presentation of his theory did not come for over 20 years, his background and deep immersion in the problem were critical to formulation of the new hypothesis.

The third stage in concept development pertains to the substitution of new conceptual schemata for old ones or the reformulation of accepted ones. Although we tend to identify major changes in outlook with individuals such as Darwin or Einstein, we must remember that this is only a type of historical shorthand. Scientific developments are the products of many minds. In the case of the evolution hypothesis, there were many prior hints, including some reasonably well developed systems preceding Darwin. By 1809 Lamarck produced a theory of evolution that could have served as a springboard for a more thorough investigation. Wallace had a theory of evolution that closely resembled Darwin's, and in fact Darwin's position was originally presented jointly with that of Wallace. There were a number of other scientists and natural philosophers of the early 19th century who had a wide variety of hypotheses about evolution.

In the "Historical Sketch" accompanying his *Origin of Species,* Darwin mentions 22 19th-century writers who had evolutionary theories of sorts in various stages of development. Why then did Darwin become the principal focus in the development of the theory of evolution? What was there about Darwin's theory that led to the rather sudden demise of scientific acceptance of the theory of special creation? Why did Darwin's presentation have such a strong impact when there were so many others in the field?

It should be recognized that one problem, perhaps the most difficult problem, in establishing a new paradigm is the attachment of scientists (and others) to the assumptions of the older paradigm.

The idea of evolution had been around for many years before Darwin. There was even a great body of data that could be interpreted as supporting evolution. Why, then, had evolution not been accepted? As Ernst Mayr points out, some of the assumptions are silent and not subject to direct verification by scientific evidence. Certain of the overt assumptions, such as essentialism, creationism, and anthropocentrism, are opposed to evolutionary doctrine. Essentialism, clearly articulated by Plato, assumes that each species has an essential nature, a fundamental principle. The idea of essentialism is strongly represented in our culture today. In essentialism, variability among category members is thought to be accidental and irrelevant. Man is made in the image of God. We have His essence; our differences are superficial. The concept of evolution requires variability within a species, and continuity between species. Creationism clearly fits with essentialism; the species must be created *de novo*. Finally, the assumption that the human species has a unique nature, probably established by divine right, keeps it from being a point on any continuum. Darwin clearly had a great deal to overcome, but he had the necessary weapons.

An accepted theory does not collapse simply from problems; though frayed and worn, it is usually clung to as tightly as Linus's security blanket. A theory collapses only when a better alternative is available. Darwin provided a better alternative. He succeeded where the others did not, partly because the time was right, but more importantly because of his presentation. He marshalled an imposing array of data to support his reasoning. And he had a mechanism—natural selection—that was more pleasing to scientists than notions of a "vital principle" or similar mystical entities proposed by some writers. The moral of this example is that an idea, no matter how good or how timely, requires substantial supporting development and evidence before it captures the allegiance that is necessary if it is to replace an accepted concept.

Our fourth point also involves the necessity of substantial development of a new concept. Despite its impact and the fact that it was more acceptable than prior concepts, Darwin's theory of natural selection was really not adequate except as an interim solution. Had there been no further development of the theory of evolution, many scientists would probably regard it with skepticism today. A great portion of this further development has come through work in genetics. Interestingly enough, Mendel's early work in genetics was presented only a few years after *Origin of Species* was published, but those who were aware of Mendel's work at the time failed to appreciate its importance to a theory of evolution. Now it is generally accepted that mutations, arbitrary changes in DNA molecules and breaks and shifts in parts of chromosomes, become part of the genetic makeup of an organism and may be transmitted to its offspring, possibly making them more adaptive to their environment than their parents were. To summarize this point, an important concept is not a static thing. New evidence and new ideas will continue to crop up; the concept must incorporate these unforeseen and often unwelcome newcomers or it will sink quietly into senility.

The Importance of Interim or Multiple Solutions

All scientific solutions are, in a very real sense, interim solutions. The solutions we are particularly concerned with at the moment are those known to be incomplete or inadequate when they are put forth. Because there is a tendency in all humans to organize incoming information in some way, all perceived events have some kind of organization. The organization may be as simple as a collective name (such as "dog") or as complex as the General Theory of Relativity. The organization (or solution) is important because it allows us to give some meaning to the incoming information. An incomplete solution thus helps the scientist interpret some information and may point to a way for organizing additional information.

Darwin recognized the incomplete character of his work in the Introduction to *Origin of Species:*

> No one ought to feel surprise at much remaining as yet unexplained in regard to the origin of species and varieties, if he make due allowance for our profound ignorance in regard to the mutual relations of the many beings which live around us. . . . Although much remains obscure, and will long remain obscure . . . I am convinced that natural selection has been the most important, but not the exclusive, means of modification.

How is it that the concept of evolution, admittedly leaving many findings obscure or unexplained, has nevertheless been one of the most important scientific schemata to date? Darwin helped scientists in two ways: by giving them a direction for research and by identifying some of the gaps in his theory. One of the most useful functions of any scientific theory is to direct research. Darwin's theory led to much important work, particularly in those areas in which the theory itself was obviously weak.

When Mendel's work was rediscovered, it was first considered contradictory to evolutionary theory. That idea was soon vanquished and much more progress was made. The locus for transmitting hereditary factors to later generations was tentatively identified as a function of the organism's *gametes* or sex cells. In 1931 McClintock and Creighton found that the chromosome contains the physical locus of these genes. In 1953, Watson and Crick identified the structure of genetic material with its basis both for replication and for mutation. Two years earlier, in 1951, Barbara McClintock published a paper showing that chromosomes are unstable and that movement of genetic elements within and across chromosomes regularly occurs. Major chromosomal differences can develop leading to clear species differentiation. The story continues, propelled, in part, by questions that were left unanswered by Darwin's theory. The world would be a tidier place if we could be sure that there is one and only one solution to a particular problem. Unfortunately, in science we are often faced with situations where more than one solution is available. When students ask

"But which answer is right?" they are assuming that there is in fact a "right" answer.

Whether there is a "right" answer to any question cannot be answered with certainty. Scientists try to generate a theoretical structure that most satisfactorily organizes the available data. The structure that they conceive is usually as coherent and comprehensive as they can make it. Although scientists would dearly love to know the "right" answers, they are aware that this knowledge can never be obtained for certain. They are content with answers that organize data and are fruitful for further research. A good example of multiple problem solution concerns the nature of light. Is it a wave form, is it composed of particles, or are there other possibilities? The right answer for a scientist depends on his use of light. In some types of research it is more useful to assume that it is a wave form; in other research an assumption of particle form is more useful. Whether light is "really" a wave form or a particle goes unanswered.

A substantial number of problems in science do have multiple answers. When two or more answers seem to be about equally compatible with the data, scientists use convenience, simplicity, or even aesthetic values to determine which solution to accept at a particular time.

The ambiguities that are a part of science are quite evident when we consider the possibility of interim or multiple solutions. We have identified the uncertain and ambiguous nature of science on several occasions in our discussions. Such statements are true. They do not imply that science is an inferior way to answer questions. Quite the contrary; science, in spite of its problems, undoubtedly provides our best opportunity to resolve a wide range of questions. Recognition of its difficulties is part of the game.

Scientific Paradigms

If you open a science book in an area that you haven't studied, you will notice very early in the book, probably on the first page, that the author introduces terms different from those you are acquainted with. You might conclude that the author is using jargon in order to impress you. Or you might feel that he or she is attempting to overwhelm you so that you won't think the book lacks substance. Most likely these are not the author's goals. Rather, the major reason everything in the text looks so foreign and forbidding is that you are venturing into a field without knowing the *paradigm* of the science.

Like most other terms introduced in this book, *scientific paradigm* is hard to define. It refers to the total complex of a science. It includes the language, conceptual framework, theories, methods, and limits of the science. It determines which aspects of the world scientists study and the kinds of explanations they consider. Most important, it includes the way scientists see the data, laws, and theories of their science. All scientific paradigms contain these multiple elements.

The scientific paradigm is different for each science, but the paradigms usually don't conflict. This book deals primarily with overlapping aspects of the paradigms of the various sciences. Within the history of any science, however, there are occasions when different scientists see the world through conflicting paradigms. When this happens there is disagreement about some of the basic tenets of the science, and scientists use the weapons of argumentation, identification of relevant facts, intuition, and presentation of new data in an attempt to convince other scientists in their domain that one paradigm is better or more realistic than another. These conflicts show that the scientific paradigms are an integral part of the science.

The concept of scientific paradigms was first systematically introduced to the authors in a very informative book by Thomas Kuhn entitled *The Structure of Scientific Revolutions,* published in 1962. But precursors to the concept, as we interpret it, date back at least to Max Wertheimer in 1912. Kuhn has published a more recent article, "Second Thoughts on Paradigms," reexamining the idea of paradigms. In this article he recognizes the complexity of the concept and the problems with using it and suggests an alternative term, *disciplinary matrix.* For the purposes of this book, however, we will continue to use the term *paradigm.*

What happens when you observe *phenomena,* which are events and happenings in the world? Certain philosophers and psychologists have argued that you first experience "raw feels," or meaningless sensations. They would say that if you were looking at a snowstorm you would experience moving spots of white that you would then have to interpret as falling snow because of your past experience with it. This would include the history of associations you have had with snow. According to this position, if you look at Figure 5–1 you must first experience horizontal, vertical, and diagonal narrow black sensations that you must interpret as lines on a paper and then perhaps as a box. But Wertheimer and Kuhn argue that it isn't natural to have "raw feels" in our experience. An interpretation of phenomena, giving them some meaning, is an integral part of the experience. If you've never experienced snow and you see a snowstorm, you may think it's manna from heaven, or dust particles, or feathers. At the very least you would undoubtedly see little white things blowing about in your environment. It is not likely that you would simply have a sensory experience—not normally, and not without a great deal of special training.

Thus, whenever people observe a phenomenon, they do not see raw facts, which they can then interpret; instead, they see interpreted phenomena. There is a major difference between saying that someone sees a phenomenon and then interprets it and saying that someone sees interpreted phenomena. The latter view seems correct. Consider Figure 5–1. You may see a box with the corners 1 and 2 in the lower front, or you may see a box with the corners 1 and 2 in the lower back. These two ways of seeing the box conflict, because you can't see it both ways simul-

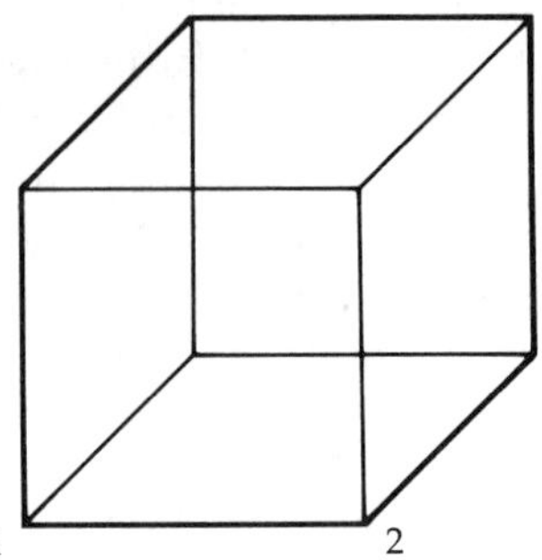

Figure 5-1.

taneously. (If you don't yet see conflicting boxes, look at any corner; if you see it as an outside corner, gaze at it intently while thinking of it as an inside corner.) The important point is that you see a *box,* rather than lines on a sheet of paper as a box. You don't see the lines as two-dimensional, because you see them in different planes. You are interpreting *while* you are observing, rather than observing and *then* interpreting. If you study this figure carefully perhaps you can see it as lines or a design on a piece of paper, but in that case you no longer see the box.

Consider Figure 5–2. Here you probably see six odd-shaped blacked-in areas. Now look at the figure intently. See anything? Look at the white background rather than the black boxes. Now that you see the word you may find it harder to see the black boxes. In like manner you don't look at a printed page and see merely spots on the paper; you see words, of course. When you look at a page containing words in a strange alphabet, you may merely see marks on a sheet of paper. That is what this page might look like to someone who cannot read our alphabet.

Now look at Figure 5–3. The bottom horizontal line looks shorter, doesn't it? And the four lines on the right don't appear to be quite vertical. However, what you see and what you can measure in these cases are quite different. The horizontal lines are indeed of equal length, and the vertical lines are parallel.

In neither Figure 5–2 nor Figure 5–3 did you actually see one thing and at the same time interpret it as something else. An interpretation is coincidental with the seeing. Of course you may later give the same phenomenon a new interpretation because you have seen it differently. But simultaneous interpretation and organization are always intrinsic to observation.

You have probably experienced new interpretations of concepts as well as of observed phenomena. As with phenomena, you see concepts somewhat differently after the new interpretation, because the observed

Figure 5-2.

elements fit a new conceptual framework. Consider the following series: O-T-T-F-F-S-S-E-N-T. What is the next element in the series? Think about it for a moment. . . . The letters seem to be in some sequence, but they don't all fit. An *O,* two *T*s, two *F*s, two *S*s but only one *E, N,* and *T.* But, now think of the numbers one to ten. Look at the series as you spell them: *O*-ne, *T*-wo, *T*-hree, *F*-our, . . ., *T*-en. The series now fits into a conceptual framework that is different from the previous one. In one sense you now *see* the letters differently, and you can continue the sequence indefinitely. The letters are part of a system, and the system now accounts for the series.

To understand the series above, you have to integrate what you know from different areas: you have to know how to count; you have to know how to spell; and you have to know how to make ordered pairs between the elements of one series and a subset (the first letters) of another. Even though you know all of these things, until you see the relation between them in a new way the series makes no sense. To learn any science (in fact, to learn any area of human endeavor), you not only have to learn what the important elements of the science are, you also have to learn to *see* these elements in relation to the other elements. It is usually much easier to learn what the elements are than to learn to see how they are related to one another. But the learning of these relations is essential if an individual wishes to add new elements to the system or to do scientific research. Knowing the scientific paradigm is essential to extending the science, in the same way that understanding the series was necessary for its extension.

We have previously discussed two domains of science—astronomy and biology—in which the paradigm changed significantly and in which important scientific discoveries would probably not have occurred in the earlier paradigms. Consider, for example, that the discovery of the planet Neptune resulted directly from observations based on Newton's extension of the Copernican theory and that the discovery of simple fossils in Pre-Cambrian formations stemmed directly from observations based on evolutionary theory.

Once an individual learns a scientific paradigm, the world of that science becomes a different world. Things are seen differently from the way they were seen before. What may have seemed central to the view before the individual learned the paradigm may now have shifted to being merely background or second-order phenomena. As in Figure 5–2, when you learned to see the white space as "THE," the black marks became

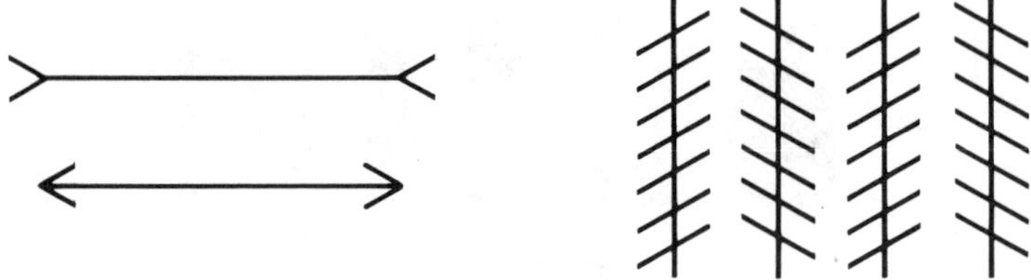

Figure 5-3.

background. In the letter series you may have noted that some letters occur late in the alphabet and others early, but when you learned the system those became very unessential facts.

Writers of science textbooks have a very difficult problem: they know the paradigm of the science, and they are attempting to teach the science to people who do not know the paradigm. What makes the task even more difficult for students is that many scientists, including those who write textbooks, don't realize that they see the world differently from those who do not know the science. Many scientists think that all they have to do is present the important facts, laws, and explanations, and if students learn these they know the science. Not so! Before they know a science, students must learn to see the facts, laws, and explanations in a certain way—they must learn a paradigm. And to learn a paradigm students must be able to see the facts and laws as being relatively natural consequences of a conceptual framework. A law must be seen to represent the way some event is supposed to act because of its nature. An observation must be an observation of some aspect of the framework.

Textbook writers will use a technical term to refer to a certain aspect of a class of events. They know that what they are referring to is very important to the problem, and it is obvious to them that the technical term is central to the idea they wish to convey. But it may be as difficult for them to realize that others don't see the importance of the ideas as it is for someone who doesn't know the paradigm to understand the importance of the term, or indeed the need for it. Scientists and students actually see the world differently.

There is one thing that aids many textbook writers in their aims: the fact that some aspects of many scientific paradigms have become part of our culture. Although the writer may have a starting point in common with the student, the overlap is not so large that the student's knowledge can be taken for granted. It would be useful if textbook writers spelled out certain of the tenets of their paradigm in an introduction to the text.

There is one aspect of scientific paradigms that needs emphasis. The paradigm is never derivable from the data, nor even from the laws. (Most subunits of paradigms are not derivable from data either.) Scientists *accept* paradigms rather than derive them. The paradigm is associated with the way scientists see a set of phenomena. There is no need for them to see the phenomena that particular way. Quite often in the history of science, different scientists have looked at many of the same phenomena and have seen them differently. Furthermore—and this is an important point—no scientific paradigm with its associated theories and explanations encompasses all the data; no paradigm is completely successful. Although the scientist may see certain phenomena in a particular way, some of the data seem to make no sense when viewed that way. The scientist hopes that slight modifications of the theories are all that are necessary to make the data fit, but there are no guarantees that only slight modifications are adequate. To know the areas in which the paradigm does and does not account for the data, the scientist must know the paradigm quite well. This is a primary reason why amateurs rarely make significant contributions to the game of science today.

A discussion of some aspects of different scientific paradigms will illustrate their necessity. Consider first classical astronomy. Children may look at the heavens at night and see spots of light. They may see them as the ancients did—that is, as a large canopy over the earth with holes in it, letting through the light on the other side. They may see the sun and the moon as traveling within the canopy, floating slightly above the mountains, and giving additional light and heat. You know that this framework has some problems. The spots on the sky move, and they do not all move together. Individuals who see the canopy with the holes in it may worry about these *anomalies* (bits of information that don't fit). They may, as Aristotle did, add a hypothesis to their theory and decide that beneath the canopy of the sky were clear spheres housing each planet. Within this framework they might ask questions such as "How high is the sky?"

Modern astronomers see the sky as a vastness beyond compare. There are in it local bodies such as the sun, moon, and planets. The sun is seen to be a huge ball of gaseous material in a constant state of activity generating a large quantity of energy. The stars are seen to be similar in kind to the sun except much farther away. The planets are solid bodies, like the earth, that travel around the sun. The clusters of stars are not organized in patterns established by the gods; they are seen in clusters only because of their position in relation to the viewer. The Milky Way is seen to be composed of millions of stars too far away to be individually identified with the naked eye. The stars are neither stationary in the sky nor moving on a canopy; each moves in its own way. Although each is moving quite rapidly, the distance between them is such that the motion is negligible in relation to the other stars.

Most of the questions that were asked about the heavens prior to this view have no meaning in the new paradigm. Everything is not accounted for here either, although astronomers can specify the problems much better than we can. Certain problems currently exist in the paradigm for astronomy concerning the origin of planets and stars and of the universe as a whole, as well as explanations of certain visual and radio phenomena.

The point is that the definitions of phenomena in the domain of astronomy have changed drastically since Aristotle's time, principally because of the new paradigm. Before the new paradigm was accepted, one early scientist, Giordano Bruno (1548–1600), anticipated the idea of an unlimited universe with other earthlike bodies circling other suns. He was burned at the stake for such heretical ideas.

Most natural phenomena are part of the domain of more than one science. Practitioners in the several sciences abstract parts of the phenomena differently because they view them in terms of different paradigms. The textbooks of these sciences point out the specific abstractions, but they fail to point out explicitly the paradigms related to their abstractions. Outlining a paradigm requires discussing phenomena at an entirely different level of abstraction—a level that is usually ignored.

If you peruse a textbook in physics you will find data abstracted in terms of measurement, vectors, forces, acceleration, stresses and strains, pressures, work, transverse waves, electric fields, charge, dielectrics, resistance, magnetic fields, resonance, amplification, refraction, images, lenses, diffraction, radioactivity, fission, fusion, and many others. The concepts implied by these terms are elaborated upon, experimental conditions to measure what they represent are described, and probably the relation of some of these concepts to the everyday world is discussed. But only rarely will the general conceptualization and paradigms be discussed.

Consider what might happen as you attempt to learn a new and strange language. You could memorize every word and still not be able to speak the language. You must also learn the rules, structure, and relationships within the language. The rules are learned slowly and painfully, but they eventually become automatic, giving meaning to messages in the language. Scientific paradigms are like languages in many ways, since all the elements are not immediately obvious and many steps are necessary before all the elements are tied together. If you want to realize how difficult it is to explain a paradigm, attempt to explain to someone the rules, structure, and relationships of something you know very well—for example, your native language. You will undoubtedly find many obscure aspects that you can identify but cannot express. Knowing a scientific paradigm is in a large sense like knowing a language. You can't learn a language without first learning some of the vocabulary, and in like manner you can't learn a scientific paradigm without first learning some of the specific concepts and relationships. Thus it is extremely difficult to explain the paradigm in an introductory science course.

The difficulty of explaining a paradigm often goes beyond the introductory level. When Barbara McClintock proposed movable genetic elements and influences from the cell onto the activity of the genes, based on cytogenetic correlates of gross differences in the organism, most of her fellow geneticists, who functioned primarily in the biochemical realm, did not understand her.

Probably no one puts all the concepts of any science together in one large view of the world at any given time. Scientists generally see a given set of phenomena in the context of similar phenomena within their paradigm, but they don't grasp the interrelation of all phenomena at once. An analogy can be drawn with the following complex sentence: "The mathematician that the chemist that the physicist laughed at avoided solved the problem." All the parts fit together, but one can't see them all at once. To understand the sentence one must analyze the different subject/object relations.

Differences among Paradigms

There are different sciences simply because each deals with different topics. The particular set of topics dealt with in a given science is known as the *domain* of that science. In some broad fashion we can say that physics started with questions pertaining to the motion of objects and their influence on one another. The paradigms of physics developed through attempts to explain how these phenomena came about. Similarly, paradigms of chemistry developed through attempts to find the essence of matter. Chemists asked "What are things made of?"; and some of the alchemists attempted to change one substance to another. Biology probably began with primitive contact with and curiosity about animals, plants, and illnesses. Psychology had its beginnings in curiosity about the minds and behaviors of organisms.

Since early scientists were interested in different things, or different aspects of the same thing, they noticed different relationships, organized different sets of facts, created different conceptual frameworks, and developed different paradigms. As we saw in the illustrations earlier, the same phenomena are actually seen differently when viewed in different contexts.

To see how the paradigms of several different sciences might be associated with a single event, let's consider a person eating spaghetti (although no scientist is likely to select this topic to study, since objects of study are also part of the paradigm and since scientists within the different sciences select only the facts that are appropriate and socially acceptable for them to study). Assume that a physicist, a chemist, a biologist, a psychologist, and an anthropologist all view this situation according to their respective paradigms.

The physicist may view the diner's fork as a rigid body that is being used as a lever. She may be interested in the problem of the spaghetti slipping off the fork and may note that the force of friction tends to be less than the sum of the projection of the gravitational force tangential to the plane of the fork and the inertia of the spaghetti. The chemist may note that the diner is eating a starch, a homopolysaccharide that results in glucose when completely hydrolyzed by acids. He may appreciate the fact that the glucose is a ready source of energy for the diner. The biologist may classify the spaghetti, stating that it comes from a particular wheat, or comment on the fact that swallowing is coordinated with a lifting of the tongue and that it initiates peristalsis. The psychologist might note that through experience the diner has learned to manipulate the fork and spoon deftly (although sometimes the spaghetti slips off, causing measurable frustration), and that, due to conditioning, the diner salivates even before tasting the spaghetti. The psychologist may also notice the expansion of the pupil of the diner's eye, indicating enjoyment of the spaghetti. The anthropologist may note that spaghetti is a culturally acceptable food in certain Western subcultures and is an integral part of certain social functions (such as spaghetti dinners).

Obviously, each scientist embeds a subset of the facts into a conceptualization accepted by the particular scientific subculture. And each scientist emphasizes his or her own set of facts. This illustration does not imply that there are not similarities or overlap among paradigms. Most of this book—in fact the very existence of this book—indicates that the several sciences have many common features. But depending on the paradigm in which an event is viewed, the different facts of the events take on greater or lesser importance. No paradigm explains all the facts, and within the paradigm some facts are more important than others. Each culture, subculture, and each person within the culture pays attention to some facts and ignores others. Some facts are more central to one view of the world than to others.

Consider yourself. You probably don't know (without looking) whether your ring finger is longer than your forefinger. You probably don't know whether you put your left sock or right sock on first or even whether you put one on first consistently. You probably *do* know the color of your eyes and how tall you are. You probably know the name of the book you are reading but not the names of the authors. Your age is an important fact about you if you want to drive a car, but your grade-point average is not. Your grade-point average is important if you want to win a scholarship, but whether you drive a car is not. The relative importance of facts depends on the particular situation.

If you are interested in where someone lives, knowing the house number is probably more important than knowing whether there is a large yard. Even though the city, the street, and the house number are all necessary to find the place where someone lives, the *order* of establishing these facts is important. The street is meaningless if the city is unknown, and the number is not very helpful if the street is unknown. An exception to this sequence may occur if you have a different paradigm for organizing the data, such as knowing that some street names have environmental restrictions. One is not likely to find a Jefferson Davis Street in New York, a William Tecumseh Sherman Street in Atlanta, or an Ayatollah Khomeini

Parkway in Washington, D.C. Moreover, one might know that a city has certain street numbers in one part of town or certain architectural styles grouped by districts.

In summary, we have seen that different people viewing the same event from different paradigms emphasize different aspects of the event; they put these aspects into different contexts; they interpret some aspects as more important than others; and they see the event differently from one another. Moreover, different paradigms lead to different organizations of data.

Scientific Revolutions

Individual scientists may puzzle over a phenomenon they do not understand; but, even without understanding it, they will give it some interpretation—though perhaps the interpretation is superficial or unsatisfactory. Even a statement such as "it is out there" for a flash of light is an interpretation of some sort, because the flash could be caused by the eye or brain. The first time an event is observed, it is given a location and probably associated with other events.

When scientists study a set of events, they try to fit as much of the data as they can into a conceptual framework. If they can accept the conceptual framework as part of a paradigm, they use this paradigm as the frame of reference from which they see the data. That is, they see the events as interpreted by the scientific paradigm.

Usually (these days, just about always) the conceptual framework and paradigm that the fledgling scientists use are those they learned in the textbooks and classrooms of their science. Students acquire the paradigm gradually by being told that certain phenomena go together until they finally see the data in terms of the paradigm.

The popular assumption that science develops in an orderly, systematic fashion is orderly, wishful thinking. Such a theory assumes that scientists go out and gather data and then find by induction the laws from which the data are deduced; a few such laws lead to a more general law. This ideal simple sequence has been proposed as the actual process of science. But in reality science is quite different.

Scientists see facts in a framework; certain facts cannot be made to fit this framework easily. Scientists usually try to extend their interpretation or reinterpret the facts so that they will fit it. Sometimes facts that seem to be central to the conceptual framework are not easily derivable from it. In this case the scientists usually work very hard at fitting the data to the framework—perhaps by adding new (ad hoc) assumptions or by gathering new data that they hope will lend support to the framework. However, sometimes after much work of this type scientists realize that the theory on which their conceptual framework is based is unwieldy and unconvincing.

The younger members of the science are often the first to appreciate the unwieldy character of the theory, although one does not have to be young to change theoretical positions. When the theory becomes unconvincing, most scientists try to modify their theory. A very few might question the conceptual framework itself, on which the theory is based. We saw an example of this process in our discussion of the theory of special creation of species and the succeeding theory of evolution. Here the notion of what is a species and what is creation was at stake. Thus, scientists sometimes propose a new conceptual framework encompassing the data that had produced the anomaly in the old one. When there are conflicting ways of viewing the data within a single scientific domain, a revolution is in progress. Scientists entering the field may be shown both conceptual frameworks, in which case they have alternate ways of viewing the data. When this has happened in the physical sciences, scientists have usually agreed after a period of time that one framework organizes the data better than the other. Either the old paradigm has continued or the new paradigm has become the way most scientists view the domain. However, in the social sciences, no one framework has ever been accepted by nearly all the scientists. In a sense, revolutions are continually in progress in the social sciences because no paradigm dominates. Social scientists thus are faced with making a choice among conceptual frameworks. Psychologists have their various paradigms, such as behaviorism and psychoanalysis; sociologists have their role theorists and institutionalists; and economists their Keynesians and classicists. There are many possible reasons for the lack of paradigm agreement among social scientists. We personally believe that the real reason is that social science is very difficult and complex. There are so many interactive factors that one may organize data in many different ways depending on how they are approached.

When two individuals in a given science look at the same event from different paradigms, they see different facts as being central to the science. In a sense they see past each other. Authors of introductory textbooks in the social sciences often present some of their facts within one paradigm and other facts within another. A conflicting presentation tends to cause confusion for the student, who thereby gets an inconsistent view of the field.

Because anomalies exist in all scientific paradigms, the door is constantly open for new proposals. Scientists do not often respond to the anomaly with a new conceptual framework. Regardless of the anomaly, most scientists tend to believe that the framework they hold is essentially correct, so they have no desire to overthrow the paradigm. Only a few unlearned or brave souls feel compelled to challenge the structure that underlies their science. Usually the challenge is met and defeated. It is a rare occurrence when a revolution is successful—that is, when a paradigm shift actually takes place. But when a successful revolution does occur, it is of great importance.

Consider an important aspect of a popular paradigm of a few centuries ago. The demise of this paradigm occurred only within the last century. The concept: The universe is completely filled. There is no truly empty space. If something is moved, the space that occurs is immediately filled by something else. This conceptualization was not discussed much but was a framework in which scientists and philosophers thought about and saw the world. It had been put into words with Aristotle's statement "Nature abhors a vacuum," which meant that nature does not allow open space; open space is really filled with air. Certain phenomena are derivable from such a paradigm. One of these is an explanation of how a drinking straw (or any suction pump) works. The drinker sucks the air out of the straw; since there can be no vacuum, the air is replaced by the liquid into which the straw is submerged. Undoubtedly, many individuals covered the far end of a straw, sucked, and found either that they couldn't draw anything out or else that by sucking they collapsed the straw. These facts are consistent with Aristotle's statement, but the following one was not. It was known that suction pumps could draw water only about 34 feet and had to be used in tandem to draw water above this height. This limitation is not really compatible with the filled-universe hypothesis, but no one formally questioned its implications until Galileo.

Galileo explained the limitation of the suction pump by an ad hoc hypothesis suggesting that a column of water broke at about 34 feet and therefore could not be sucked up beyond that. The same idea can be illustrated by stretching a wad of chewing gum: you reach a point where it breaks. You may have tried to stretch a drop of water over a short distance by touching it with a point and pulling it; you will notice that it is quite difficult, because the drop soon breaks. Galileo suggested that water would break under its own weight at about 34 feet and that 34 feet was therefore the pump's limit.

Torricelli, a student of Galileo, became interested in this phenomenon, and for some reason he didn't accept Galileo's explanations. He hypothesized a new conceptual framework in which the reason water rose in a suction pump was that we are in a sea of air and the air pushes down on the water. If the air is removed from above one spot, the air pushing down on the rest of the water would force it into the area where the air was removed. You can see the same phenomenon by putting a boat in a tub of water. The boat forces the water down beneath it and the water compensates by rising around it. Torricelli's explanation of why water would rise only 34 feet was analogous to the boat. The boat displaces only as much water as it weighs. The water that rises in the pipe rises only until its weight becomes equal to the weight of the air pressing down on the outside of the pipe. Above the raised water, Torricelli thought, we have a vacuum! He tested his theory by using mercury. Since mercury is 13.6 times as dense as water, Torricelli expected it to rise much less, only about 30 inches. He filled a tube that was longer than 30 inches with mercury and inverted it in an open bowl of mercury. The tube did not remain filled, but rather the mercury fell until it was about 30 inches above the level in the bowl. Although Torricelli confirmed his expectations, others still saw the space as filled. They claimed mercury broke after 30 inches and water after 34 feet because invisible bonds could only support a certain amount of weight. Pascal, in turn, showed that the mercury would not rise as high in the tube if the experiment were done at a higher altitude where the air above it weighed less, but still people believed that all space was filled. Even after Boyle showed that a liquid, which was raised in an inverted tube in a chamber, dropped as the air was pumped out of the chamber, some scientists *still* held that nature abhors a vacuum. They may have believed that Torricelli, Pascal, and Boyle's results were simply anomalies that needed to be resolved in the context of the plenum (the filled universe). According to later scientists, space was filled with an ether that was weightless, colorless, and odorless and could flow around airtight stoppers. Existence of the ether was partially based on the idea that waves need a medium to travel through, and light waves travel through space. But the rise and decline of the ether theory is another story.

The change in the view of the content of the universe from being filled to being partially empty had vast consequences. It was a successful revolution. Individuals following Torricelli, Pascal, and Boyle now saw that much of the atmosphere was open space and that weight and pressure were actually the factors in many situations in which the antagonism of nature to vacuums had been the earlier explanation. Revolutions such as this one are unpredictable in both occurrence and outcome and can be either upsetting or rewarding, but they are an intrinsic part of the game of science.

This chapter has dealt with two main themes. First, ideas do not stand in isolation. Their development depends on both personal and social circumstances. Second, ideas are held by individuals. It is the way that individuals interpret ideas and the events associated with them that constitutes a paradigm. Scientists spend many working hours attempting to show that the paradigm within which the events are seen enables one to understand those events. If it does not account for important data, scientists try to introduce modifications so that it does. Sometimes different scientists find themselves looking at the same data differently. If the new way of looking at the data seems to fit, a battle for the minds of the scientists takes place.

6
Scientific Inquiry

How is scientific information gathered? What are the relevant considerations for such operations? When is a scientific theory confirmed or negated? What are the properties of a scientific theory? What are the reasons for accepting one theory and rejecting another? These questions are basic to the process of scientific inquiry. Some of the answers have been suggested in the previous chapters. In this chapter we shall give them more attention.

In Chapter 5, scientific paradigms were illustrated. The discussion pointed out that everyone sees the world from one point of view or another. People may switch paradigms at times, but they always operate within one. The general limits and properties of the paradigm are usually implicit in scientific behavior. Scientists rarely attempt to make them explicit. In general the ideas and concepts held by scientists seem to them to be consistent with the paradigm, even though many aspects of the paradigm are either loosely defined or not defined at all. Scientists usually consider only a small set of concepts within a paradigm and attempt to articulate, in detail, the operations and interactions of those concepts.

The scientific enterprise takes place at several levels of specificity. The most encompassing level is the paradigm itself. Scientists have some ideas of the kinds of data that are relevant to their science and some general way of interpreting the data. They also have some ideas of the kinds of laws that are likely to relate the data and the kinds of mechanisms that are acceptable as explanatory devices. They "know" what kind of research is acceptable and which experimental designs to use. They also "know" which aspects of different situations are relevant and which are irrelevant. They have ways of evaluating what is good research and what is not, above and beyond the actual experimental designs employed. They will evaluate some research as appropriate and some as inappropriate, even when all the research falls within the official domain of their science. If you ask "Appropriate to what?" the implicit answer is "Appropriate to the paradigm." Thus the formal domain of the science is not necessarily the actual domain.

Within the paradigm and sometimes considered coextensive with it is what we have been calling the conceptual framework. This term refers to the kinds of functional relations used in describing the domain of the science and the coherent framework that structures the way the observations hang together. The techniques of investigation and the definition of good and bad research are not directly part of the conceptual framework, although they are part of the paradigm. The paradigm sets constraints on the kinds of theory and experiment that scientists accept as being valid. Scientists consider only theories and data that are deemed consistent with these constraints. The paradigm can be thought of as spectacles that color what scientists look at and blind them to alternative conceptions and "irrelevant" data.

Scientific Theory

If we consider the hierarchy of units within the scientific enterprise, we find that the paradigm and conceptual framework are more encompassing than the theories but are also generally not spelled out by the scientist. Narrower in scope than theories are specific theoretical and empirical concepts, laws, experimental methods, and data. Theories are the level at which scientists consciously explain the phenomena of their domain.

Use of the Term *Theory*

Before continuing, we should look at the difference between the scientist's and the layman's use of the term *theory*. You've heard people say "That's just theory; it's not a fact." Or "It works all right in theory but not in practice." Such statements show a great misinterpretation of the term *theory*. *Theory* in science does not mean an unconfirmed statement. There may be very strongly confirmed theories as well as unconfirmed ones. In fact, scientists are not likely to call a statement a theory until they have quite a bit of confirmation for it. They will use the term *hypothesis* or *tentative theory* until they are relatively certain that they are correct. If we wish to claim that something is unimportant because it is only a theory, then it is unimportant that the earth is round, that electricity flows through wires, and that hydrogen nuclei combine to form helium and give off energy in doing so.

It is nonsense to say that something is correct in theory but not in practice. A scientific theory is usually defined in terms of an idealized set of conditions; it may be that a particular situation outside of the laboratory to which the theory is applied is so complex that the application is tenuous, but theories *are* about the real world. Either they are sound in practice or they aren't sound at all. A theory is the basic symbolic system developed to account for observations, and a good theory is one that accounts for or explains a great number of observations simply and concisely.

Quite frequently the theory and the paradigm cannot be clearly differentiated. When scientists think about a domain, they may not separate what is written from their "intuitive feel" of the material. The paradigm, containing the conceptual framework, refers to the way scientists look at their domain. It is not simply a set of interrelated sentences, and it is not something that is easily affirmed or, in particular, easily falsified. One does not reject a paradigm on the basis of a single prediction that is not supported by the data. Instead one modifies the details of the theoretical articulation of the framework while keeping its core concepts inviolate. An unsupported prediction is simply an anomaly to be resolved in the framework. This can be done by searching for the causes of the anomaly or by minor modifications of the theory involved. Thomas Kuhn called this process of resolving "puzzles" the process of normal science. It is only when the anomaly is repeatedly investigated and no satisfactory resolution is found that some scientists may begin to challenge the paradigm.

Some paradigmatic ideas include Aristotle's idea that everything in the universe has its natural place and Galileo's idea that all spatial frames are equivalent. "All behavior has a well-defined cause" is a paradigmatic notion, as is the idea that life systems can be explained by the same kinds of laws as inanimate systems. Relativity and quantum mechanics concern themselves with the paradigmatic ideas of fields and atoms.

Certain paradigms are not considered scientific, in part because their supporters make no attempt at generating accompanying theory. Special creation and a belief in psychic phenomena are two such paradigms. In spite of comments to the contrary by Ronald Reagan while campaigning for the presidency in 1980, evolution is not simply a theory, but a broad-based, multiply supported way of viewing a great variety of phenomena and events in a coherent, systematic way, and it supports a variety of interrelated theories. It is thus a paradigm.

Relation between Theory and Experiment

Even though we write about the relationship between theory and experiment, we realize that many sciences depend in part or almost exclusively on field or observational studies. Consider some: evolution, astronomy, plate tectonics. In each case, organized observation takes the role played by experiment elsewhere. Data are organized and explained, theories generated, and predictions made and tested. The techniques differ; the general principles are the same. Sometimes experiments and field studies play complementary roles. In the case of animal behavior, phenomena such as migration are studied in both field and lab, but it would take quite a lab to study the social behavior of blue whales.

Scientific observation requires identifying an observational context and selecting certain aspects within that context to observe. An experiment merely formalizes that operation in certain situations. In what follows, our primary descriptions are based on experiments, even though the ideas discussed refer to both experimental and nonexperimental research.

The purpose of a theory is to describe and explain observable and observed events and to predict what will be observed under certain specified conditions. In order to test a theory, scientists may conduct experiments by setting up conditions for observation. It is a step toward confirmation of a theory if the events observed in the experiment are predicted by the theory, and it is a step toward rejection of the theory if the events observed are contrary to those predicted by the theory. Thus an experiment provides the situation in which specific events can be observed, and the results of a successful experiment tend to confirm or negate a theory.

Sometimes a theory does not state precisely what the relations between the elements are. In fact most if not all theories are incomplete. That is, they have necessarily left at least some aspects of the relations between elements undetermined. For that reason, some experiments may be designed to provide data that will help scientists define certain relationships among the theoretical elements. For example, in the development of theories of supersonic flight, experiments were designed to investigate the relationship between the shape of an airplane and the amount of friction in the air at and beyond the speed of sound. Prior to these experiments, theories specified the relationship only up to the speed of sound.

Experiments are sometimes designed for another purpose. Sometimes there are two or more competing theories covering the same domain of science. In such cases an experiment may be designed to pit the two theories against each other. If the two theories predict different observations in the same situation, the scientist sets up this particular situation so that the observations can be made. The theory that predicted the actual observations seen tends to be confirmed, and the one that did not predict the observations correctly tends to be negated. (Many scientists may prefer the term *hypotheses* to *theories* in this context. There is, however, no formal difference between these terms. Generally if an hypothesis has been confirmed to some degree and a group of scientists believe it to be true, they are likely to call it a theory.)

There may be experiments, however, in which the same results are predicted by both theories. In such cases, the theories are said to be *equivalent*. If there is no *possible* way to observe events that would differentiate between the two theories—if no experiment can differentiate between them—the theories are said to be *logically* equivalent.

An example of two equivalent theories revolves around the experiment of dropping two rocks at the same time. A popular Greek theoretical explanation was that rocks and earth are made of the same element and different members of this element like to be with each other. The theory stated that if a rock is released it goes toward the earth because it cannot bear to be away from it. As the rock gets closer, it gets so excited that it travels faster. This theory may be equivalent to Newton's theory of universal gravitation: every particle in the universe attracts every other particle with a force directed along the line joining them, proportional in magnitude to the product of their masses and to the inverse square of their distance of separation. Some modifications of the Greek theory to include the statement that all particles like each other proportionately to their masses would make the two theories logically equivalent.

When scientists attempt to present a new theory, they may have a difficult time getting others to pay attention to it. However, if they can design a certain kind of experiment—sometimes called an *experimentum crucis,* or *crucial experiment*—they may have considerable effect on other scientists. In this kind of experiment the theory makes a "psychologically unique" prediction—one that is "obviously" not consistent with the prevailing paradigmatic assumptions and thus seemingly could not have been predicted by any other theory. Such an experiment, if successful, tends to give the theory more status than do experiments with predictions that seemingly could have been made from theories within the current paradigm. It is important to note, however, that a *psychologically* unique prediction does not imply a *logically* unique one. No prediction can be logically unique, because there are always an unlimited number of possible theories from which an event can be predicted, even if people know of only one theory. Furthermore, even theories within the prevailing paradigm can almost always be modified to account for any strange result. Psychologically unique experimental results have much more effect on the development of a science than others, although there is no logical reason for this disparity.

Consider the following example concerning the generation of lower organisms. A controversy long existed on how certain individual organisms are formed. Some scientists thought that these creatures are spontaneously generated—that they are produced by some "vital principle" or "virtue of the infusion." Other scientists thought that one living organism of a species has to produce another. Scientists on both sides did much research, but the opposition remained unconvinced. The usual experimental technique required that substances be boiled in bottles to kill the organisms in a culture. If new organisms appeared, one side argued that the bottle's cork was porous or that all the organisms had not been killed. If no new organisms appeared, the other side argued that the boiling destroyed the virtue of the infusion for generation of life or the virtue of the air for supporting life. During the early 1860s, Louis Pasteur sterilized a culture in specially designed flasks with long slender necks bent at sharp angles. Time passed but no organisms appeared. This seemed to end the controversy. It seemed that only the theory that one organism is necessary to produce another could account for these results. However, now one can easily see that the dead air in the flask could have blocked the passage of the vital principle. Incidentally, the case for spontaneous generation was reopened in 1876. In the final analysis, Pasteur's "crucial experiment," like all crucial experiments, was one experiment among many.

There is another important function of a theory as it relates to experimental research. This function has to do with the actual design of experiments. One of the functions of a good theory is that it sets up the conditions for meaningful experimentation. A person who doesn't work within the framework of a theory has no basis on which to design experiments. If you have taken a course in an experimental science, you may have wondered at the stupidity or dullness of some of the experiments. They are probably not stupid or dull to everyone. The meaning of the experiment depends on the scientific paradigm and the theories to which it relates. If you don't understand the paradigm of a science, you very likely will not see the meaning of the experiment. If you understand only the methodological aspects of a theory, you can perform a designated experiment by observing specified procedures, but you aren't likely to design a meaningful experiment yourself.

Many people believe that an experiment must always give some sort of an answer. Once a venerable dean, when told that a young research worker had run a series of studies but had gotten no meaningful results, asked whether scientists aren't supposed to follow the data wherever they may lead and then publish the results. But the truth of the matter is that many studies, particularly in basic research, don't give definitive answers and may actually be uninterpretable. Most scientists have cabinets, boxes, and wastebaskets full of data that can't be interpreted. Why? There is a saying that, if anything can go wrong in a study, it will—and at precisely the wrong time. The list of things that can go wrong is infinitely long: concepts may be wrong, the conceptual framework may be inappropriate, the logical sequence from the theory to the observation—which is always complex—may be incomplete, the wrong measurement may be made, equipment may falter, an assistant may prove inept, variables may not be appropriately controlled, an error in design may be overlooked, the sample being measured or tested may be atypical, there may be an unknown variable, or who knows what. In any case, the results may not make any sense.

When an experiment *is* interpretable, it is always interpretable within some theoretical framework. The fact that a meaningful interpretation of data requires working within the framework of a theory is missed by some of the people who write about science. They portray science as a series of blunders, a stumbling by accident from one important discovery to another. Although some discoveries do occur by accident, most do not. Scientists prepare carefully and observe carefully. Unexpected events do occur, but scientists who always chase unexpected events whenever they occur will make no progress. One reason why scientists need a theory to test is so that they will be able to identify those events that are predicted and those events that are not predicted as they proceed. Careful scientists spend a great deal of time gathering and evaluating data to develop a theory that is detailed and definite enough to reveal, during an experiment, any event that is antagonistic to the theory. If they have no program of research with theories and expectations, they cannot recognize the unusual. The history of science is filled with important discoveries being made and going unnoticed several times before a theory is ready to capitalize on them. If an unexpected event occurs but is unrelated to the current experimental plan, careful scientists almost never drop everything else to follow up the unexpected event, although they are likely to keep that result in mind for possible future investigation. (Perhaps one of the reasons that accidental discoveries are played up so much is that the name that has been given to accidental discovery, "serendipity," sounds so poetic.)

What is an unexpected event? It is one that does not fit naturally into the conceptual framework that researchers use to understand the observations. If the framework is not reasonably complete, enough so that it "ought" to function in a certain way, there can be no such thing as an unexpected event and no real chance of conceptual advance. Scientists must have a conceptual framework in order to know whether the results meaningfully support their theory or refute it.

The research scientist normally performs a series of interrelated experiments. Experimental investigations are the football linemen of the game of science; they don't gain much publicity, but the science cannot advance without them. A single experiment cannot advance scientific knowledge very much, but it can fill a hole in a theory or help open the scientist's mind to a new insight. In any case, a good experiment will allow the experimenter to ask additional questions.

To understand some of the uses of scientific theory and to discuss in some detail experimental methodology, let's look at a detailed example.

An Example of a Scientific Theory

As our example, we can use the conditioning theory of human behavior. First of all, we should look at some of the aspects of the paradigm in which the theory exists.

This particular paradigm of human behavior developed within the scientific milieu of the 19th century. Scientists had begun to make predictions more precisely than they had ever dreamed possible; an understanding of the natural world seemed within their grasp. Some laws within physics and chemistry were seen to be equivalent, and even certain aspects of living organisms were found to conform to laws of physics. Biological scientists thought that they could reduce the complexities of life to the laws and theories of physics and chemistry. This thought received a boost with the acceptance of the theory of evolution, which proposed that more complex organisms develop from less complex ones. With the synthesis of organic compounds from inorganic materials, scientists could conclude, and did conclude, that all living organisms originally developed from inorganic compounds and thus that the laws of chemistry and physics could account for all life processes. This interpretation of the laws was accepted as part of the paradigm.

The paradigm, in general, has in it the belief that one should explain any complex activity as simply as possible. Thus simple laws should in turn be explained by simpler ones. In this paradigm, known technically as *reductionism,* complex behaviors such as social interaction could be explained by the laws of the behavior of single individuals; in turn, the laws of individual behavior could be explained by the laws of physiological activity, physiological activity could be explained by the chemical activity within the organism, and complex chemical activity could be explained by simple chemical action. Since the paradigm holds that simple chemical action is at base responsive to the laws of physics, its premise is, in effect, that all activity, including social interaction, is ultimately derivable from the laws of physics.

Within the behavioral sciences, particularly psychology, another application of reductionism became part of the paradigm. According to this position, complex behaviors are explainable by the laws of simple behavior. Also, the same basic laws of behavior hold for all organisms. If the laws of behavior are explained by physiological processes, it seems reasonable that the laws of behavior are generalizable among the species of animals—especially since the paradigm assumes that species are related to each other. Thus some psychologists have as part of their scientific paradigm the assumption that the same laws of behavior hold for all animals under all conditions. They assume that there may be some quantitative differences among animals and conditions but no qualitative ones. We should note here that many psychologists accept some aspects of the conditioning paradigm as we describe it but not all aspects, and the concepts of sociobiology are in some ways radically opposed to it (see Chapter 7).

Working within the paradigm of reductionism, the Russian physiologist Ivan Pavlov became interested in the process of digestion. Pavlov hypothesized that salivation and chewing are *reflexes* caused by food on the tongue. When food reaches a certain consistency resulting from chewing and salivation, it will cause an animal to swallow; the swallowing will begin muscle activity called *peristalsis,* which carries food to the stomach. There the food will cause glands to secrete acid and enzymes, and the process will continue. According to Pavlov, one event will always lead to another until the process is completed.

To summarize the situation as it existed for Pavlov: He viewed the world through the paradigm of reductionism, believing that complex behaviors are reducible to simple ones and that they are all ultimately reducible to laws of physics. He thought that laws of digestion are the same regardless of the species and that all complex behaviors can be reduced to their component parts, which are reflexes. He demonstrated experimentally one instance of this paradigm by developing a theory of digestion. Using dogs as his experimental animals, he put them in harnesses so that he could control their behavior and operated on them so that he could observe the processes he was interested in. Carefully controlling events by stimulating the animals at different phases of the digestive process, he studied their reflexes. During this investigation Pavlov noted an unexpected phenomenon: at certain times he got erratic behavior—as, for example, when the animal salivated *before* food was put on the tongue.

Pavlov did not immediately conclude, as some would suppose, that the occurrence of salivation prior to physical stimulation proved the inadequacy of his theory of digestion. He did not drop everything to investigate this unexpected event. The accidental discovery didn't lead him to change his ways or his paradigm. He continued his work on digestive reflexes. Only after he came to some conclusion on his current line of research—enough to be awarded the Nobel Prize in 1903—did he investigate "psychic secretion," as he called anticipatory salivation.

Since Pavlov's paradigm included the concept of reflex behavior, he decided to find out what event led to psychic secretion. Pavlov reasoned that if the animal salivated before receiving food because of a reflex, then some event prior to the salivation must have initiated that response. Thus salivation must have been caused by some change in the animal's environment.

Before continuing with the discussion of Pavlov's research, we need to digress for a bit in order to consider variables and experimental design. These concepts have been discussed on a common-sense level earlier in the book, but the specific kinds of variables Pavlov used and the conditions under which they were observed are more easily understood from a more formal point of view.

Variables

Almost all discussions of experimental design and scientific methods either explicitly use the term *variable* or imply the concept. A variable refers to a specific aspect of events or things. It differs according to the particular event being considered, and no single event or thing can have more than one value of the same variable at a given time. The concept can best be understood through examples. The weight of a person is a variable: different people weigh different amounts, and the same person weighs only one amount at any given time. The number of times you blink in an hour is also a variable; in any given hour you will blink a certain number of times, and this number may differ from hour to hour.

Variables are not necessarily personal. The color of cars, the make of cars, the number of cylinders in a car, how far the car goes on a gallon of gasoline, how much the car cost—all are variables. The concept represented by the term *variable* is a very pervasive one. Whenever an entity or event is described, every dimension by which it could be described can be considered a variable, including where it is and its environmental conditions.

One of the most important things scientists do when they conduct an experiment is to attempt to find out what the relationship is among different variables.

Let's consider what is meant by a relationship between variables. How fast can you run 100 yards? It is obvious from what has been said that the time it takes to run 100 yards is a variable. It is a variable that can be *quantified.* To be quantified simply means that the different values the variable takes are established by some method of measurement. Quantitative values are usually numbers. But how fast can you run 100 yards? What if you were running on grass? What if you were running uphill? Or after a large Thanksgiving dinner? What if you were wearing hip boots? What if it were raining? Such things as the kind of track, the slope of the ground, the time since you last ate, the kind of clothes that you are wearing, and the weather conditions are all variables. They all probably have some effect on your running speed. The color of the clothes you are wearing, the depth of solid rock beneath the track, the time since you last went to the movies, and the number of sisters you have are also variables. These latter variables are unlikely to have much effect on running speed; there is a relationship, however, between the kind of clothes you are wearing and running speed, because the heavier your clothes the slower you run. Handicappers use this relationship between variables in an attempt to equalize a horse race; the horses that are known to run faster have to carry more weight.

If it is true that the more weight people carry the slower they run, we can say that there is a *functional relation* between weight carried and running speed, or we can say that running speed is a function of weight carried. If we wished to state this relationship in symbolic terms we might do it this way: let S equal running speed and W equal weight carried; then $S = f(W)$. This equation is read as "S equals f of W," or "S is a function of W," or "Speed is a function of weight carried." A more technical way to state this part of experimentation is to say that scientists investigate the functional relationships between variables. They study the systematic changes in one variable as another variable changes. The scientists may simply be interested in establishing the general nature of the functional relation, such as "The more weight one carries the slower one runs." If they are interested in establishing a specific relationship between variables, they might do so by constructing a verbal description, a graphic illustration, or a mathematical equation.

Experimental Design

What are the problems encountered when a scientist conducts an experiment? In designing the experiment he or she has to be very careful to see that the relationships found do exist between the variables being investigated. He or she must try to hold *constant* the influence of variables other than those being investigated.

In an experiment, there are three kinds of variables: (1) independent variables, (2) dependent variables, and (3) control variables.

Independent variables. Independent variables are those whose values are directly manipulated by the experimenter. The experimenter is interested in establishing functional relationships between independent and dependent variables. If, for example, an experimenter is interested in determining the relationship between weight carried and running speed, he can do so by getting someone to run 100 yards on a number of different occasions. The runner can carry a pack each time she runs, and the experimenter can add some rocks each time. The runner may run six times, carrying either 0, 10, 20, 30, 40, or 50 pounds of weight on each run. Since the experimenter can directly manipulate the amount of weight carried, weight carried is an independent variable.

Dependent variables. Dependent variables are measures of events that are the product of the experimental conditions. Whereas the values of the independent variable are in a sense assigned before the experiment begins, that control is not possible with dependent variables. Dependent variables are those that are affected by the actual process of the experiment. In our current example the dependent variable is the time it takes the runner to run 100 yards. This time is measured under each of the values of the independent variable. If the experimenter wants to state the results in the form of a functional relation, he says that running speed (the dependent variable) is a function of weight carried (the independent variable).

Control variables. Control variables are crucial. They are the conditions that make an experiment an experiment. These are variables that the experimenter does not want to vary systematically with the independent variable. In other words, the average value of the control variable should not change as the independent variable changes. The experimenter thus sets up experiments so that these control variables cannot systematically affect the relationship between the independent and dependent variables. The major problems in experimentation are associated with control variables. Consider our example. If we are investigating the relationship between running speed and weight carried we do not want other variables to influence running speed. We would want the race to be run under as similar conditions as possible for each of the values of the independent variable. The runner should be well trained before the experiment begins so that she does not improve with practice during these six occasions. The races should be run on successive days so that the runner's energy will not decrease with each race. She should run all the races alone, or all of them with someone else, so that the desire to compete will be about the same each time. She should run each time on the same track so that the composition of the track will not affect the outcome. She should run each race at the same time of day (allowing the same amount of time to elapse after eating), so that the weight of the food and the physiology of digestion will not have different effects on the outcome of the race.

In fact, to be more sure of the relation between running speed and weight carried, the runner should probably carry each weight more than once. There are so many possible influences on the time it takes to run that the experimenter cannot possibly control them all. Say the runner runs with the same six weights on many occasions. She runs once a day for three months, with the conditions as nearly the same every day as the experimenter can manage, except for the weight carried. On successive days she carries different weights, and the same weights are carried regularly throughout the period; then different influences—such as changes in humidity and temperature, which the experimenter may not be able to control, and the feelings and attitudes and temporary physical conditions of the runner, which the experimenter certainly cannot control—will tend to average out over the different values of the independent variable so that they will not systematically influence the final result.

Another way to attempt to control the influence of other variables in finding the functional relation between weight carried and running speed is to give to each of a number of different people the same weight to carry and have each run 100 yards, and then to give other people different amounts to carry, until many people have run with each weight. In this way the influence of all the other variables will again tend to average out. Some people will be more tired than others, some in better shape than others, and some more highly motivated than others, but hopefully some of all kinds will run in all conditions. The results under these conditions will not be very precise for any individual, but they may be precise enough to show a relationship between the independent and the dependent variable, and the experiment will take less time than the first method.

One problem that this example highlights is very often involved in experiments—particularly in the social and biological sciences. It is that certain variables that are likely to have an effect on the relation between the dependent and the independent variables cannot be made the same for all tests. For example, when different people are assigned to carry each weight, the experimenter must be aware of the possibility that those carrying, say, the 40-pound weight have all just finished dinner, while those carrying the 10-pound weight have just gotten out of bed. The experimenter can control for this by using a technique for which there is clear mathematical justification, called *random assignment*. In this technique the participants in the experiment are randomly assigned to a particular value of the independent variable. Thus the people carrying the 40-pound weight will be selected from among the participants by some random procedure (the drawing of straws, the pulling of names from a hat, a table of random numbers, and so forth). In fact, all assignments of objects or participants to be tested or used in any experiment should be random. When that cannot be done—for example, when the independent variable is a naturally occurring one—one must be *very* wary of not being able to control certain variables that may vary systematically with the dependent variable. This may frequently occur when an experiment is done and the independent variable is sex or race or socioeconomic status. Consider the difficulty of interpreting IQ tests given to people of different cultural backgrounds and social roles. Are the differing results due to the social situation or the genetic makeup of the participants?

This question, of course, has become very controversial. People are very interested in assigning the source of the differences in IQ between groups to hereditary causes or to environmental causes, depending on particular political beliefs. This tendency is so strong that anyone who jumps into the fray is going to be called names. Since the independent variable in these studies is not randomly assigned to members of the same population, one can present arguments against any conclusions drawn.

To be sure we understand the important aspects of basic experimental design, let's briefly consider one other hypothetical experiment.

Consider the relation between the amount of a substance dissolved in water and its boiling point. The independent variable here is the amount of a substance dissolved—such as 1, 2, 3, 4, 5, or 6 teaspoons of salt. The dependent variable is the temperature of the water when it boils. Control variables are such things as the amount and source of the water, the consistency of the salt being dissolved (the salt for all values of the independent variable should come from the same well-mixed source), the kind of container, the temperature and humidity of the room, and so on.

In summary, an experiment contains three primary elements:

1. *Independent variables,* which are directly manipulated by the experimenter so that each experimental condition has a particular value of each.
2. *Dependent variables,* which are measures taken during the experimental process.
3. *Control variables,* which are variables that should *not* vary systematically from condition to condition. This can be achieved in three ways: (a) Have no variation of the control variable among the different experimental conditions. (b) Have the same distribution of the control variable among the different experimental conditions. (c) Have the values of the control variable randomly assigned among the different experimental conditions.

The experimental scientist has the task of trying to relate some terms of a theory to dependent variables and some of them to independent variables. Then, if the scientist can design an experiment that permits observation under controlled conditions, he or she will find out what the relationship is between the independent and dependent variables and, through them, what the relationship is between the theoretical terms. If the theory makes predictions, the experiment can tend to confirm or negate it. If the theory is not explicit, the experiment can give the evidence necessary to make it more explicit.

An Example of the Relation between Experiment and Theory

Now let's return to Pavlov and the example of conditioning. Pavlov was interested in finding out how events that preceded food on the tongue came to elicit salivation. He thought that such erratic behavior was due to reflexes of the higher nervous system. Pavlov also thought he could find the conditions that caused this erratic performance. To investigate it, he first had to define a dependent variable and an independent variable. Salivation was an obvious dependent variable. Since Pavlov believed that some event preceding it led to the salivation, he varied the presentation of a cue before salivation. His independent variable was whether or not a metronome sounded before food was placed on the tongue. Many controls were necessary to do the experiment. Pavlov used dogs as a control variable so that he would not obtain variable effects from different species. He put the dogs in a harness and in a room with extraneous variables held constant. For example, the room was kept at a constant temperature and humidity. It was both soundproof and lightproof. Food was delivered directly to the mouths of all animals without any physical disturbances in the room. The dogs were in the same state of hunger for all conditions.

Pavlov found that the first time the clicking sound of the metronome came on, the dogs perked up their ears and looked toward the sound; then they received the food, which caused salivation. The same sequence of events was repeated. After a very few pairings of the click and the food, some of the animals started salivating before the food was presented. Pavlov had rediscovered psychic secretions; this time, however, they did not seem mysterious at all. Now he knew that the salivation *followed* the click, even though it preceded the food. Salivation was a response to the click.

Pavlov had demonstrated that the complex behavior of getting ready to receive food could be explained by procedures he was familiar with. In fact, it was no more than a special kind of reflex. Salivation due to the food itself is called an *unconditioned reflex;* salivation due to the click is called a *conditioned reflex.* The difference is that the unconditioned reflex appears whenever the food is presented, whereas the conditioned reflex would not occur without a history of past associations between the food and the click.

Pavlov's later investigations indicated that any "neutral stimulus" (such as clicks, lights, or buzzers) could be associated with an unconditioned reflex and then elicit the response of that reflex. Conditioned reflexes can thus be established by pairing neutral stimuli with unconditioned reflexes. The conditioned reflexes tend to disappear when the neutral stimuli are presented alone. Many other laws were developed, and conditioning, as the technique came to be called, became one of the fundamental procedures in the study of learning. The techniques and results would not have been found if Pavlov had not formulated a theory with which to interpret the events.

More recent research in learning has led to a great number of specific additions and changes to the theory. Some psychologists think that the research has led to modifications that are awkward, perhaps contradictory, and untenable. If the theory is rejected it will be because detailed experimentation on anomalies has shown its inadequacy and has led to a body of data that can be interpreted more simply through a different theory. Thus, even a faulty theory serves a major purpose: without it, better theories cannot be developed.

Observational and Survey Methods of Science

There are other ways of gaining scientific information besides controlled experimentation. These methods tend to be neither as powerful nor as efficient as experiments; however, there are many occasions when experiments cannot be run to attain the data sought. Sometimes scientists may go to the area in which events that they are interested in are likely to occur and then either simply observe, carefully making records of the events observed, or else employ instruments to record events they might not notice with the naked eye. Closely related to these direct-observation techniques are techniques in which some variable is sampled in several places or times. This kind of repetitive observation is called a *survey*.

Scientists cannot control the earth and the heavens, but they can observe them. They can, for example, take field trips to record heavenly events during the time of a solar eclipse. This was done in 1919 and 1922 to get data relating to Einstein's theory of relativity. These expeditions gave evidence that light does not always travel in a Euclidean straight line but that it bends as it passes near a large mass, such as the sun. Geology, geography, biology, sociology, anthropology, psychology, archaeology, political science, and history are all sciences that must rely on observational and survey techniques for some of their data. An attempt to get an understanding of complex social or historical events is likely to require observational or survey techniques rather than straight experimentation.

Although he is an astronomer, Gerald Hawkins used observational techniques to make a major archaeological discovery. About 3900 years ago, some inhabitants of Britain started to assemble a huge arrangement of stones, which, with the mounds and holes around it, is known as Stonehenge. All the stones—the largest of which was about 30 feet long and weighed about 50 tons—had to be transported many miles. Some stones weighing over 5 tons had to be transported 250 miles. But the most interesting discovery made by Hawkins is that these stones and holes near them were set up to be a complex astronomical observatory and computer. The "observatory" identified the locus and possible times of solar and lunar eclipses and the locus of the sun and moon at significant times of the year—for instance, solstices and harvest moons. The occurrence of significant events is quite irregular, so these prehistoric people included counting devices to predict when they would occur. Hawkins took careful measurements from Stonehenge and used an electronic computer utilizing current knowledge to compare modern estimates of when the events actually occurred with the measurements the ancient Britons could have taken. The agreement between the Stonehenge measurements and the occurrences of heavenly events as computed from current knowledge is extremely high; there is almost no likelihood that it is accidental. Agreement between theory (Stonehenge was an astronomical computer used by prehistoric people) and observation (the angles measured point to places where significant astronomical events occurred) was confirmed. Thus we now believe that ancient Britons had a civilization that was highly organized and included much division of labor, great technical skill, and detailed astronomical knowledge. This ancient civilization was organized and coherent enough to work on, use, and expand this architectural masterpiece for at least 400 years.

In psychology and the social sciences, some scientists are interested in particular characteristics of different individuals or groups as they interact with their natural habitat. What is the means by which a particular primitive society distributes important materials to its members? What are the child-rearing practices in the society? How varied are they? What are the particular social conditions that tend to lead to the greatest hostility toward minority members of our society? How do the institutions of the society—for example, the banks and the courts—differentially treat members of different subgroups in the society? What are the major issues that affect the voting patterns of a particular community? These and many other questions are the kinds that are invariably answered by first going into the field, observing, asking questions, and checking the reliability of the data collected.

Such phenomena as differences among people, the ways in which people behave in different complex social and environmental conditions, or the effect of governmental policies on behavior must be studied in their natural surroundings with whatever analytic and observational tools are available. The task is difficult, but the results can be quite illuminating. In collecting data, social scientists must, above all, try to collect the most appropriate and unbiased data that can be obtained and then be very careful about drawing conclusions from that data. They must always worry about bias and the possibility of alternative explanations. However, if they know their topic well and are continually aware of the possibility of bias, they are less likely to draw unsupportable conclusions.

In research the general problem is to work in such a way that we know what is affecting the result we observe. We speak of supportable and insupportable conclusions. What makes a conclusion supportable? Supportable conclusions in the scientific enterprise come from empirical research done in the laboratory or in the field. The first step in gaining supportable conclusions is finding out what it is that affects the phenomena observed. That is, scientists first make observations. If the situation were different, the phenomena they observe would not be the same. They try to find out what in the situation makes a difference and in what way the observable phenomena would be different if the situation were to change. In laboratory research scientists learn what affects their observations by controlling some factors (control variables) and manipulating others (independent variables). The situation can be clear, very complex, or indeterminate in observational studies. An example of a clear situation can be seen in astronomy. Here we can reasonably know the factors at work (mass, relative movement, and so on). In field work like that often done by sociologists or animal-behavior researchers, variations in such factors as genetic structure and life experiences may be unknown, and other factors, such as status and behavior of significant others, may vary conjointly. The researchers' confidence in the results varies accordingly.

Since it fits here about as well as anywhere, we will discuss the use of animal models. Both of the authors were interested in learning and one was a confirmed rat runner for many years. Students and others often ask "But what does this have to do with humans?" Sometimes a great deal; sometimes nothing. You don't know until you try it on humans. The same can be said no matter which two species or subspecies are being considered. Behavior modification, which is currently being widely used in areas such as the treatment of behavioral disorders or problems and acquisition of skills, originated in the study of rats and pigeons. You have an Rh blood factor. The Rh stands for rhesus, since it was first discovered in rhesus monkeys.

In practice nonhuman animals are used in behavioral and physiological studies for a number of reasons. Among these are cost, knowledge of genetic background, better control, and reliability. By reliability we mean that the subjects are always there and sober when needed. There are also problems of ethics involved. If a scientist wants to learn about the effects of removing a chunk of the brain, few humans would volunteer; nonhumans aren't consulted. Open-heart surgery was developed using dogs. Rhesus monkeys were used to develop polio vaccines. Some people object to the use of nonhuman animals for experiments, particularly those experiments that disable or cause pain. While there are governmental rules for the use of nonhuman animals, this ethical problem still rests with the researcher.

In brief, then, scientists build models of behavior or biological function using nonhumans. Whether the models fit humans is a problem for further research.

Decision Processes in Science

We have discussed both experimental and nonexperimental methods for the acquisition of scientific knowledge. We have not yet discussed in a single location how one evaluates the evidence for a theory or hypothesis. The task of evidence evaluation is, like most of the other issues in this book, not well defined, although many scientists seem to think that it is.

There is a standard procedure that most science students learn in methodology classes, which helps them make scientific decisions. One of the authors actually teaches such courses. According to the "received" view, decisions are determined in science by the outcomes of statistical tests. But statistical tests are only one part of the decision process. These tests are primarily applied to the numbers generated by an experiment.

In an experiment, a scientist sets up several situations that are almost, but not exactly, the same. More than one measure is generated by repeating each situation several times. These repetitions give the numbers for the tests. One question a statistical test often helps to answer is: Are the numbers from the different situations similar enough to each other to be thought of as being from the same set? If they are, the experimenter concludes that the different situations did not make any difference in whatever they measured.

Let's return to the example mentioned earlier when we were discussing the relationship between variables. Suppose one collected data on how fast people run carrying different weights on their backs. Ten randomly selected men from a gym class may be given zero pounds to carry, and groups of ten others 10, 20, 30, 40, or 50 pounds each. Each weight defines one situation. Each of the ten men in a group is a repetition. The time it took each to run 100 yards would be measured to an accuracy of tenths of a second. These are the numbers. An analysis of variance (one of many statistical tests) may be run on these times. This test helps answer the question "Are the running times associated with each weight essentially the same?" If the distribution of running times with the different weights are mathematically similar to one another the answer would be yes. If they are different, the answer would be no. If the conclusion is that they are different, the scientist may try to describe the mathematical relation between running times and weight carried.

This experiment can be seen to be very limited empirically. Only men from one class from one school were in the experiment. Women, children, other classes, other schools, other nations, and other species were not represented. The weights were of a specified set, probably carried in a particular way, and so forth.

Other experiments can be run. Depending on the theory being evaluated, these experiments may be very different from one another. It could be a theory of movement quickness, a theory of stress, or a theory of balance and stability. Different sets of data would become involved depending on the theories tested. However, the relationship of the different data sets to one another and to the theories is determined by analyses logically distinct from the standard decision processes. This would be true regardless of whether other experiments were to be run or other theories were to be evaluated. The analyses are made by using these data as part of an argument in defense of the theory.

Experiments are usually designed to fit into some statistical model so that probability statements can be drawn of a type similar to the example just cited. Hopefully, a reader of this book can now see that the statistical result is only a small component of experimental design and an even smaller part of the scientific enterprise in general. As far as the research is concerned, one needs to have a reason to run this particular experiment. How does it fit in with the concepts and ideas of interest? Will results of the experiment allow the theory to be better articulated? If they come out one way, will they supply evidence that would tend to confirm the theory, or will the experiments tend to refute the theory if they come out in another way?

The standard decision process does not help answer any of these questions. All it does is help the scientist specify what the relation is among the different samples of numbers derived from the different experimental groups. Any other data derived from any other source has to be integrated into the theory by logical or informal argument, which lies outside of this decision process.

The standard decision process requires the structuring of data collection according to very strong constraints so that the data can be entered into some previously described mathematical structure. When these constraints can be applied without distorting the conceptual system to which it is applied, the formal nature of the decision process allows the scientist to be quite precise about the validity of that decision. The world, however, is not always so well ordered. Even when an experiment is run, the interesting data often cannot be entered into the formal decision process. Barbara McClintock, for example, is one of the most careful observers in the world. Many of the structural relations she saw between the observations of chromosomes and the observations of corn were individual occurrences. She found many different occurrences of different patterns of relations. The theory she proposed accurately accounted for relationships under many different patterns. Generally, she went directly to the integration of ideas without doing very many statistical tests.

When one does observational research, it is important to relate the observations to theoretical concepts and structures. When one can say that one observation of a set of relationships illustrates a theoretical pattern, several other observations are consistent with the theoretical expectations. Further, if one can claim with justification that if the observations were different in certain ways they would tend to refute the theory, the data could then be seen to support the conceptualizations. These conclusions may be reached without requiring either experiment or statistical decision.

For instance, there are at least two ways to talk about the future in English. Do they mean different things? What is the difference? The two ways we have in mind is the use of *gonna* (going to) and *will*. Julie Gerhardt studied this problem among children. She observed three or four pairs of three-year-olds playing different games. She also observed the context in which they used the different terms and the social consequences of doing so. She noticed that *will* (or *'ll*) was almost always used to refer to events that were either ongoing or to be immediately carried out. When *will* was used there was almost always agreement between the children, and they were usually playing together. When *gonna* was uttered, the described activity often was not carried out. *Gonna* was used when the children disagreed, or when they were doing different things, or to describe quite unlikely events far in the future.

Gerhardt used these distinctions and others to argue that the two terms under study entered into two different thought and social patterns. *Will* helped signify cohesion and sharing between the children, and rather than being descriptive of the future, it tended to mark ongoing activity. *Gonna* represented more individual planning. Neither actually signified the abstract future.

All scientific research rests on the regularity of the data as defined by the conceptual system. Standard decision processes are just some of the tools by which one finds the regularities in the data. Researchers in artificial intelligence represent components of theories in computer programs. They input data into the computer and then see whether the outputs are coherent according to their conceptual system. For example, one can check whether a theory of good chess playing is valid by whether the system plays a coherent game and whether it wins or not.

Like McClintock, Gerhardt carefully observed the relationships among certain phenomena. The data that her theory tends to account for is not of the type that can be clearly categorized into predefined mathematical formulations. The goal of the research is to find a theoretical description that coherently accounts for the data collected. In all science this process goes beyond the statistical decisions. Regularity is found in the way data are accounted for by the conceptual structure. By rationally relating the conceptual system to the data, determining whether the data fits that system, and modifying the theoretical concepts if the data do not fit, the process meets the scientific criterion of being self-testing and self-correcting.

Formal Characteristics of Science

We have discussed some of the factors that are involved in scientific inquiry, and we have seen that empirical results provide data used by a scientist to confirm, develop, or modify a theory. Now we should briefly discuss some formal concepts involved in the development and confirmation of theories.

To have a scientific theory is to have beliefs about a certain domain of the natural world. It is a conceptualization of how that aspect of reality specifically functions. The theory is about objects and events, some of which may be easily observed, some of which may be observed only indirectly and by devious means, and some of which may not be observable at all. The belief in nonobservable objects and events is justified by assuming that their consequences are observable.

One well-known theory that well illustrates the employment of unobserved entities and events is the thermodynamic theory of gases. It postulates free molecules moving about within some specified volume. The pressure is a result of the average number of molecules colliding with the walls of the region per unit area, and the temperature is a function of the average speed of the molecules.

For a theory to have any permanence it has to be given some form other than a belief. It needs to be told to someone else or written down. It needs to be expressed in some form. The usual form of expression is to use sentences of a natural language, such as English, possibly in combination with mathematics. Though not usually considered part of a theory, diagrams, physical models, or pictures may be employed to clarify the theory. How the expression of the theory relates to the phenomena that the theory is meant to explain is a difficult problem. We will present some of the ideas that have been proposed to solve it. This will help give you a feeling for why it is difficult to formally relate language and experience.

One of the issues that scientists and philosophers of science worry about is precision. They want scientific theories to be accurate and precise. One might think all the terms used in the sentences of the theories should be clearly and accurately defined. What happens if we try to define all of the terms? A "term" is a word or a phrase. Definitions are given in words. The words used to define the first set of words also need to be precise. Define them. Their definitions also will be in words. One can see that regardless of how many words are defined, either there will be more to define, or some of the words that were previously defined will be used in the definitions. The problem can be made more real if you consider whether you could learn a foreign language from a dictionary in that language. Every word you looked up would be defined with words that you did not know until you looked them up. You would thus be trapped in the dictionary.

One solution, which was proposed by Rudolf Carnap, among others, was to consider some of the terms as "undefined" and define all terms but the undefined ones. If the selection of undefined terms is made judiciously, the definition sequence would finally come to rest with all other words in the theory defined by combinations of the undefined terms. One problem with this idea is that it is very hard to implement. Such a process was never done for any meaningful science. It is not clear that anyone could make up a set of undefined terms that would work. Even if someone could come up with a theory with the appropriate format, he or she would then have to give it an observational base.

There are several proposed solutions to this problem. One might take the undefined terms and give each of them a meaning. One way is to give them a *coordinating* or an *ostensive* definition. That is, they could be defined in terms of some sensory experience. In this way one could define words such as *blue, round, hard, pound* (weight), *meter,* and the like. It's not clear, however, how all theories could be reduced to such terms.

There have been other proposals in this kind of analysis. A physicist, Percy Bridgman, proposed *operational definitions* for theoretical terms or sentences; for example, the *length* of an object is the number of times a meter stick could be laid end to end without going past it, or intelligence is the score on an intelligence test. No one has ever shown how any theory can be constructed by using these methods, but the issue as to how to clearly specify the meaning of a theory is a real one.

What is included in a theory? What kinds of sentences are there? One needs sentences that describe the properties of the theoretical entities. For example, molecules will be considered to be like tiny, perfect billiard balls—perfect in the sense that they would bounce better than superballs; if dropped three feet, they would rebound three feet. One also needs sentences that express the relation between theoretical concepts, such as $F=ma$ (Force equals mass times acceleration). Finally, it is necessary to have sentences that observations may possibly show to be either true or false. If there are no sentences of this type, the theory will have no empirical value. (This is a formal statement of the falsification doctrine described in Chapter 4.)

What is the nature of a theory? It is possible to define a theory as the set of sentences by which it is expressed. It is also possible to define a theory as the meanings of those sentences. Since there is no satisfactory explanation of what a sentence means, under this definition theories are ill-defined. In either case, one has no formally valid way of showing that a theory is clear and meaningful. One has to rely on indirect methods to reach this conclusion. One of these methods is to tell other people what the theory is and have them show that they understand it. Do they claim to understand it? Do they make the same predictions from the theory? Do they describe events in similar ways?

We like to think of a scientific theory as a conceptual model. It consists of a representation of the objects and the events in terms of the properties that the theorist believes those objects have. The sentences, equations, or drawings that are used to represent it or to communicate it are designed to allow the individuals communicated with to build a conceptual model that matches that of the theorist. Another function of some of the sentences of a theory is to allow someone to derive predictions without actually learning the theory. One might even have a computer derive predictions.

In this chapter we explored the process of scientific inquiry. We find that this inquiry has many informal guidelines but very few well-defined principles. In order to be a scientist one must relate observable phenomena to some conceptual system. One cannot make this relationship in a haphazard manner but must show that the conceptual system can be related to more than one situation in a consistent and coherent manner. Moreover, it must be the case that the scientist has some means of using the conceptual system to limit the set of possible observations that would be considered acceptable.

7

Some New Developments in Science

Games are not without a beginning. They also wax and wane in popularity, divide into parts, and change their appearance and rules. So it is with science. In this chapter we will look at the birth and growing pains of three of the new games in town.

When most people think of science they are most likely to think of the traditional natural sciences such as physics, chemistry, and biology. Others might include the social sciences such as anthropology, economics, psychology, and sociology. In reality these sciences have become superstructures that encompass a variety of disciplines. Within each, fields have emerged that a newly arrived little green man would have difficulty seeing as part of the same discipline. Ethology is one of these. It bears some birthmarks from zoology but has a sufficiently distinct character to justify its independent identification. In addition, new sciences are developing that cut across three or more sciences.

At this time, we shall discuss two of the new, hybrid sciences and briefly comment on a third. Many such hybrids have appeared in recent years—for example, cognitive science, genetic engineering, bioengineering, sociobiology, program evaluation, and paleogeology. We wish to explore the beginnings and development of sociobiology and program evaluation. Although these share the characteristic of being hybrids, they are quite different in many other aspects.

We chose sociobiology for several reasons: first, one of the authors is interested in the area; second, sociobiology may be the focal point of a scientific revolution; third, the idea has provoked an interesting scientific and nonscientific debate; fourth, the debate has led to an examination of some crucial ideas and problems in science; and fifth, the history and future development of this idea are both important and they allow you to see an unresolved movement in science.

Sociobiology

In 1975 Edward O. Wilson's *Sociobiology: The New Synthesis* was published. The response was immediate, widespread, varied, and explosively emotional. Some scientists and nonscientists were lavish in praise, while others characterized it as at least fraudulent and perhaps even fascistic. The book was discussed on the front page of the *New York Times* on May 28, 1975. Other articles quickly followed in a variety of popular publications, including Boston's *Sunday Globe, People, House and Garden, Mother Jones, Time, Psychology Today,* the *National Observer,* and *Business Week.* It is doubtful that any scientific book or idea has stirred so much popular interest since Darwin's *Origin of Species.*

Background and History

Before we consider the idea and the controversy surrounding *Sociobiology*, a little background and history are in order. Prior to the publication of it Wilson had been known primarily as a world authority on the behavior of ants. His book *The Insect Societies* (1971) is considered a classic by both foes and admirers. *Sociobiology* is a rather large book in all senses of the word. There are 663 pages, the book is 10 in. by 10 in., and it contains over 2600 (we estimated) citations. If printed in ordinary book size (happily, it wasn't), it would run over 1000 pages—a veritable *War and Peace* of scientific books. Recently, a shorter version has been published.

Except for the first two chapters, which set out some general philosophy and definitions, *Sociobiology* is primarily concerned with nonhuman behavior. The final chapter, "Man: From Sociobiology to Sociology," involves the application of sociobiological ideas to human social behavior. Although this chapter is only 28 pages long, it has received more attention than the other 635 pages from both nonscientists and scientists. It is interesting that Darwin in *Origin of Species* (sixth edition) devoted only one paragraph to humans, yet it was also a focal point in the controversies that followed:

> In the future I see open fields for far more important researches. Psychology will be securely based on the foundation already well laid by Mr. Herbert Spencer, that of the necessary acquirement of each mental power and capacity by gradation. Much light will be thrown on the origin of man and his history.

More on this parallel later.

The central idea of sociobiology was neatly rendered by David Barash: "This is the heart of sociobiology: The application of evolutionary biology to the social behavior of animals, including *Homo sapiens*." Wilson's definition seems a bit broader: "the systematic study of the biological basis of all social behaviors."

In practice, scientific fields define themselves over a period of time by the type of work done and the direction that the work takes, not by explicit definition. In any case, sociobiology seems destined to be a combination of biological principles and data applied to social phenomena. As such it touches on or may encompass such conceptual domains as behavioral biology, neurophysiology, behavioral ecology, population biology, ethology, behavior genetics, comparative and social psychology, sociology, and parts of anthropology. A hybrid, indeed.

As with any "new" scientific idea, sociobiology has a rather extensive background and history. Gerald Holton found some of the beginnings in Thales of Miletus, Anaximander, Heraclitus, Democritus, and Anaxagoras, all Greek philosophers and materialists. Skipping nimbly over the next 2400 years we come to a renewal of the idea of a physical explanation of biological phenomena by Karl Ludwig, Emil du Bois-Reymond, Ernst Brucke, and Hermann von Helmholtz. They joined a battle against vitalism, the notion that, in addition to the physical and chemical laws, unique forces or vital spirits were critical in life processes. Brucke and du Bois-Reymond stated the contrary view clearly: "No other forces than common physical-chemical ones are active within the organism."

The clearest antecedent of sociobiology is Darwin, who was not unaware of the relationship between evolution and some behaviors. In the introduction to *The Expression of the Emotions in Man and Animals* (1872) he wrote: "He who admits on general grounds that the structure and habits of all animals have been gradually evolved, will look at the whole subject of Expression in a new and interesting light." If we substituted "social behavior" for "Expression," this statement could fit into the introduction to any book on sociobiology.

In retrospect, it is a bit puzzling why the analysis of behavior in terms of evolution was so long delayed. It's true that the Social Darwinists gave evolution a bad name by their very broad claims that "success" implies a genetic superiority. One of the better known quotations that expresses this notion comes from John D. Rockefeller, Sr.: "The growth of a large business is merely a survival of the fittest. . . . It is merely the working out of a law of nature and a law of God." But Social Darwinism probably doesn't entirely explain why the evolution of behavior received such slight attention. Perhaps the early uncritical application of the ideas of evolution to behavior, as in George Romanes' book *Animal Intelligence* (1882), frightened off would-be contributors. Some of Romanes' good points, such as his discussion of curiosity, were ignored for about 50 years. In any case, with notable exceptions such as Jacques Loeb, C. O. Whitman, William James, and Oskar Heinroth, there was little detailed consideration of the relation between evolution and behavior until the early 1900s. Even then, biologists largely ignored behavior, psychologists concentrated on learning, and sociologists followed Emile Durkheim's idea of the autonomy of the study of society. The learning psychologists' work was derived from evolutionary theory, but they did not investigate how selection leads to particular and varied behaviors. A few hardy souls in psychology such as Karl Lashley considered the evolutionary framework, but their numbers and impact were limited.

Ethology emerged in Europe in the 1930s and came to the United States after World War II. Calvin Beer described its earlier phases: "To oversimplify . . . for the purpose of making a point, classical ethology could be described as Neo-Darwinism applied to behavior. . . . It also followed Darwin in construing the study of behavior as the study of instinct." Note that while ethology has some common elements with sociobiology, its emphasis on instinct is quite different. Ethologists view behavior as being more rigidly controlled and stereotyped than do sociobiologists.

It is clear that the sources of influence on the ideas of sociobiology were many and varied. We have left out many more than we have included. Our apologies.

Characteristics of Sociobiology

According to Wilson, the term *sociobiology* was used by John Paul Scott in 1946. In 1950 Scott suggested *sociobiology* as a term for the "interdisciplinary science which lies between the fields of biology (particularly ecology and physiology) and psychology and sociology." Between 1950 and 1970 the term *sociobiology* appeared in a number of journal articles, but the terms *biosociology* and *animal sociology* were also used. In 1971 Wilson titled the last chapter in *The Insect Societies* "The Prospect for a United Sociobiology."

The study of social behavior has long been dominated by those who consider the explanations of that behavior to lie in individual experience and particular environmental conditions. If you look at textbooks in sociology and social psychology, you will find few discussions of possible biological and hereditary factors related to social behaviors. In most mentions of possible hereditary factors, the comments range from slighting to outright derision. It might also be noted that most social psychologists, sociologists, and cultural anthropologists have an extremely limited knowledge of nonhuman social behavior. We hope that this situation is being changed, at least in some programs. Although it is not the whole story, part of the negative attitude of social psychologists and sociologists toward biological and hereditary factors in behavior stems from a long-standing suspicion or fear of a hereditary elite.

Most proponents of sociobiology come from a tradition different from that of the sociologists, social psychologists, and cultural anthropologists. Sociobiologists' backgrounds are generally in the study of nonhuman behavior. Thus they may feel free to concentrate on the behaviors as they exist, without any external criteria for what behaviors may be desirable from some philosophical position.

To get an idea of how some sociobiologists might approach a problem, let's consider the relationship of human adults to infants. Emphasis might be placed on those aspects of adult responses to infants that are common to human groups. One such common element is that infants are attractive to adults of *both* sexes. This characteristic is shared by most other highly developed primates (orangutans seem to be an exception). There are some local and situational variations in expressions of this human attraction (as there are in monkey troops of the same species); nevertheless, the basic relationship remains. Here, then, is a critical element in many types of social organization. If this element is considered solely from the standpoint of the environment and culture, you may get one sort of answer. If, as the sociobiologists suggest, you view *both* hereditary factors *and* culture and environment, you may get a modified or even different answer to the basis of similarities and differences in the relationship of adults to infants.

When *Sociobiology* was published in 1975, it was greeted with warm applause, violent denunciation, and calm analysis. For the purposes of our case study we will concentrate on the last two items and some rejoinders.

Opposition to Sociobiology

In introducing *The Sociobiology Debates,* Arthur Caplan stated the importance of sociobiology better than we could:

> Independent of one's personal views of the adequacy of sociobiology as a scientific theory and of the validity of claims concerning the relevances of sociobiology for the study of human behavior, and human nature, it is of vital importance to realize that sociobiology stands as an instance of a rarely observed intellectual phenomenon: the attempt to produce and legitimize a new scientific discipline. The birth pangs of sociobiology provide an invaluable opportunity to study the complex variables that enter into the creation of a new scientific field and its subsequent acceptance or rejection both by other scientists and by the general public. Personalities, empirical evidence, ideologies, explanatory power, theoretical elegance, intellectual tendencies and all the other factors commonly supposed to be involved in the legitimization of theories in science are all plainly manifest in the reception which greeted the appearance of Wilson's book. Sociobiology is of interest not only for its actual claims and possible implications, but as a case study in understanding the scientific enterprise itself.

The most violent of the organized attacks on sociobiology came from a group called the Sociobiology Study Group of Science for the People. The group included some well-known scientists as well as students and a variety of others. Although neither of the authors has had direct contact with the group, it seems fair from the written record to see it as having a strongly political approach to science. One of the opening shots in the campaign against *Sociobiology* appeared in *The New York Review of Books* (November 13, 1975) and was signed by 16 individuals associated with Science for the People. A few quotations give the flavor of the political and personal attack:

> These theories [biological determinism] provided an important basis for the enactment of sterilization laws and restrictive immigration laws by the United States between 1910 and 1930 and also for the eugenics policies which led to the establishment of gas chambers in Nazi Germany.
>
> We think that this information has little relevance to human behavior, and the supposedly objective, scientific approach in reality conceals political assumptions. In his attempt to graft speculation about human behavior onto a biological core Wilson uses a number of strategies and sleights of hand which dispel any claim for logical factual continuity. What Wilson's book illustrates to us is . . . also the personal and social class prejudices of the researcher.

On December 11th, Wilson replied to the charges:

> I wish to protest the false statements and accusations that comprise the letter. . . . This letter . . . is an openly partisan attack on what the signers mistakenly conclude to be a political message in the book. Every principal assertion made in the letter is either a false statement or a distortion. On the most crucial points raised by the signers, I have said the opposite of what was claimed. . . . I felt that the actions of Allen et al. represent the kind of self-righteous

> vigilantism which not only produces falsehood but also unjustly hurts individuals and through that kind of intimidation diminishes the spirit of free inquiry and discussion crucial to the health of the intellectual community.

This phase of the debate did not end here; further comments and rejoinders followed. We have selected a few words and phrases as indicative of the tone:

> determinist. . . . a work of advocacy on a selective picture of human history. . . . misuses the basic concepts and facts of genetics and evolutionary theory. . . . invent *ad hoc* hypotheses. . . . a form of "scientific reasoning" that is untestable. . . . another related ploy—expands the realm of these weak analogies. . . . Nothing is explained because everything is explained. . . . Any investigation into the genetic control of human behaviors is bound to produce a pseudo-science that will inevitably be misused. Nothing we can know about the genetics of human behavior can have any implications for human society.

(This last sentence assumes an infallibility and knowledge of the future that strains the notion of scientific modesty.)

It seems fair to say that the nature and vehemence of the attacks distressed Wilson. In a 1976 article he mused about the possibility (as advised by some colleagues) of making no answer. Wilson wavered about giving some lectures and withdrew from at least one public talk. However, he did continue to defend himself and was joined by others, notably Nicholas Wade in *Science.*

A reasonable question can be raised as to whether attacks or the possibility of attacks such as we have described briefly inhibit research and speculation. We have considered this problem and discussed it informally with colleagues. Some indicated that areas of research such as racial and sexual differences, recombinant DNA, and even some aspects of history are too loaded with political consequences for their taste. Some hardy souls do work in highly controversial areas and in some cases, such as DNA, even reduce the emotional atmosphere. Others, however, obviously enter the arena because they are interested in notoriety, reveling in the publicity, interviews, debates, and celebrity status that it brings. Quite frequently, a great number of pages are published generating much more heat than light. Those who avoid such areas do so for a variety of reasons. Fear of consequences or a distaste for personal confrontations motivate some. Others may judge that rational answers are not possible in an irrationally emotional climate and defer any such work. They do not wish to spend the majority of their professional time in political rather than scientific effort. Whatever the position of individual scientists, research in many of these fields actually gets done; it may be delayed sometimes, but in the long run it is effective. Most scientists, however, go through a lifetime of research without encountering any such barriers. That does not mean that would-be censors are not with us and do not continue to be a burden to science.

Analysis of the Idea

Recently, the political discussion of *Sociobiology* seems to have become quieter and more restrained, which brings us to our next phase, calm analysis. In fairness it should be pointed out that not all the questions raised by the Science for the People group were personal or political. They did question the completeness and interpretation of the data.

Two authors who were not associated with the debates we have been considering have raised a number of important points. S. L. Washburn and Frank Beach are both leaders in the field of behavioral science. We cannot consider their arguments in any detail, since to do so would go far beyond the scope of this book. Summarizing very briefly, they pointed to some rather uncritical use of evidence by Wilson, followed by loose interpretation of the evidence. Washburn and Beach were not unaware of the potential of sociobiology but adopted a wait-and-see attitude.

Wilson made some rather broad claims for sociobiology. There are explicit statements that certain disciplines such as ethology and comparative psychology will be cannibalized by sociobiology and will cease to exist. When one looks at the domain of sociobiology, one can see that there are also implicit claims that it will absorb other fields, perhaps de-

velopmental psychology, social psychology, and sociology. It seems to us that some of the rather broad and far-reaching claims are premature at the least and probably wrong in their implication. Ethel Tobach chaired a symposium at the 1978 meeting of the American Psychological Association in which the status of comparative psychology was discussed in light of the sociobiology movement; the *American Psychologist* published that symposium in November 1980. Several authors pointed out that processes such as learning, perception, and motivation are not handled in detail by sociobiological concepts. In fact, such concepts seem to be rather naively articulated by sociobiologists. The symposium authors contended that sociobiology cannot become the universal science of behavior because its explanatory framework is too global; it uses major, long-term influences extending over generations to account for behavior. Many of the other sciences are more interested in the day-to-day differences in within-species behavior controlled by environmental flux. It is obvious to us that these two orientations are complementary rather than antagonistic to each other.

Wilson's claim that sociobiology will usurp the domains of many related disciplines is probably wrong. However, that is not a reason to reject the positive contributions of its central thesis: that understanding biological processes aids the understanding of social events.

Along a slightly different line David Hull, a philosopher, posed the question "Scientific Bandwagon or Traveling Medicine Show?" In a very lively essay Hull looked at two earlier scientific movements, phrenology and evolution, and viewed their trials and tribulations as similar to sociobiology's. Hull found the evidence and logic of these movements to be amorphous and evasive. He pointed to changes Darwin made or accepted in his theory, but he claimed that modern evolutionary theory is even more diffuse than Darwin's original pronouncements. He considered the fact that there is much use of ad hoc arguments in sociobiology, but he found this neither surprising nor particularly objectionable. In Hull's view, today's ad hoc hypothesis may be tomorrow's law. Hull viewed methodological problems in general, and he saw the objections raised against sociobiology as reflecting standards applicable to finished products and not to a newly emerging theory. He dismissed the idea that Darwin and Wilson were advancing dangerous ideas by asking how we decide that an idea is dangerous, thereby implying that the question is unanswerable. He viewed good "scientific method" and the scientific enterprise in general as being justified by the outcome. Phrenology did not establish itself for a number of reasons, including an emphasis on skull shape rather than on localization of brain function. Evidence never arrived to show that head shape was systematically related to intelligence or personality. Evolution was more successful because, although evolutionists differed among themselves in some ways, they produced a shared conceptual framework and a great deal of data around which they could rally. The rediscovery of Mendel's research spurred evolution's continued development. In short,

Hull, Washburn, and Beach seem to have reached similar conclusions; sociobiology must be allowed to establish its credentials rather than be rejected out of hand. We agree.

Before leaving the land of sociobiology we would like to consider its broader perspective. Sociobiology can easily be considered a conceptual framework within which scientists may develop theories, rather than as a theory itself. The framework proposes that when one wishes to develop a theory of social behavior, potential biological and evolutionary determinants should not be ignored. The framework of sociobiology clearly is not theory-neutral. Anyone working in the framework would almost certainly hold to the proposition that human and animal social behavior is influenced by its biological substratum. The sociobiologist would also believe that an animal's biology has been shaped by its evolutionary history. One would thus study social dynamics, in part, by the nature of the ecological niche in which the animal functions. To what extent do evolutionary factors influence this behavior? What are the mechanisms by which this influence occurs? These questions cannot be answered prior to the development of research and theory on the matter.

It is clear that theoretical issues surrounding sociobiology are many, and they will not be answered by one or even a few empirical studies. It requires the study of the interaction of the history of organisms, the history of species, and the local environmental press with feeding, fighting, and mating, as well as maternal and other individual and social behaviors.

The picture of sociobiology is being painted. Sociobiology is the basis for several journals and textbooks. There are courses expounding its concepts and graduate degrees being given for research in the field. Only if all of this research came up empty would it be concluded that sociobiology has failed. As of this writing, it is joining the ranks of viable sciences.

Cognitive Science

Cognitive science is concerned with the scientific study of mind, information, and intelligence. It has as its goal an understanding of how one can represent intentions, purposes, knowledge, and thinking in a physically realizable system, such as a person, animal, or computer. The discipline has had a lightning growth, suggesting that the times are ripe for such a science. Many universities already have research centers and degree programs in cognitive science. There is a Cognitive Science Society and several journals dedicated to research in cognitive science. Private, state, and federal granting agencies support research in cognitive science. Whether all of this is a temporary fad or whether cognitive science will prosper and take its place among the recognized sciences, only time will tell.

Cognitive science, like sociobiology, has antecedents that extend back throughout the history of ideas. Ideas central to cognitive science started coming together in the mid-1950s. The invention of the computer and the exploration of concepts associated with it were central to these developments. The 1950s gave us such ideas as stored programs, automatic symbol manipulation, and new procedures for arriving at decisions (heuristics). Allen Newell, Herbert Simon, Marvin Minsky, and John McCarthy are major early contributors to these developments.

The expression *cognitive science* made its appearance in the mid-1970s, and in 1979, at a meeting in San Diego, it was "founded" as an independent discipline. The disciplines from which it gets many of its ideas and methods include computer science, linguistics, philosophy, neuroscience, psychology, anthropology, and many others.

It is interesting to note that cognitive science is almost the antithesis of sociobiology in many ways. Sociobiology stresses that the evolutionary history of each organism plays a paramount role in its nature and therefore in explaining its behavior. By implication, according to sociobiology, in order to understand an organism's behavior a scientist must understand the evolution of its species. Cognitive science, on the other hand, stresses the similarities of intelligence wherever it may occur, whether it be in dolphin, monkey, man, or computer. Many cognitive scientists hold that the *nature* of the intelligent organism or machine is almost irrelevant to that intelligence itself. Intelligence can be programmed into a computer, or it may derive from evolutionary history; the origin of this intelligence is only of secondary interest. Since cognitive scientists are often more interested in the nature of intelligence than the nature of the organism that has that intelligence, some of them use a concept such as *cognitive agent* to stand for any of these possessors of intelligence.

There is a consensus among cognitive scientists that in order to demonstrate intelligent behavior a cognitive agent needs to integrate information from several different sources. In particular, information is required both from the environment and from the agent's memory, and the agent needs some processes or procedures to select, modify, and evaluate that information. For example, in order to play chess an agent must know the rules of the game and must be able to perceive where on the board the pieces are. In order to play well, the agent should be able to make legal moves and to evaluate positions in terms of their relative "goodness" value.

Although it seems obvious that an agent must know things in order to act intelligently, it is often not obvious what information needs to be known nor how to use it. Interestingly, some problems that are really easy for people to solve turn out to be very difficult to understand. Consider the following examples.

What exactly do you need to perceive, and what do you need to know in order to recognize a saltshaker on a dining room table? There is a clutter of visual information that reaches your eyes. This has to be coordinated for two eyes and stabilized over head-and-eye movements. The stabilized information has to be separated into different objects, with different components seen to belong to the same object. You must know something about the various shapes, sizes, and colors that a saltshaker may take, and you must be able to recognize them. As complex as this description is, it is still incomplete, and it doesn't bear on what is needed for you to locate the saltshaker in space, pick it up, and use it appropriately.

What information do you need in order to understand a sentence? The context within which the sentence is uttered and the perceived reason why it is uttered may affect what the sentence means. "Do you have the time?" may be a substitute for "What time is it?"; or it may be a request for help on a calculus problem. You must be able to parse the sentence into words. You must be able to retrieve the appropriate meaning and syntactic role for each of the words. You must be able to relate those meanings in the appropriate way, and you must be able to relate the meanings to the proper topic. In addition, you must be able to evaluate the validity of the sentence, modify your knowledge of the objects referred to, or respond either verbally or behaviorally in a way that is appropriate to the meaning of the sentence. You are almost certain to derive conclusions about the objects and events mentioned, conclusions that go far beyond the propositions actually expressed in the sentence.

Other questions of interest to cognitive scientists include: How do cognitive agents solve logic problems? How are faces recognized? How does an agent parse a complex scene into its component objects? How can these objects be recognized? How do experts organize their knowledge about a topic? How do they differ from novices? How does an animal know how to find its way home? How does a bird know how to smoothly set down on a limb or on the surface of the water? How does a tennis player know exactly when to swing her racquet? or where? How does a physician diagnose an illness? What the agent knows, what it needs to know, and how it uses that information is the domain of the cognitive scientist.

Cognitive science and sociobiology are not necessarily in conflict with each other. Even if they considered the same phenomena they would tend to relate them to different data bases. (Remember the spaghetti example of Chapter 5.) Both new disciplines are currently generating interest both in the scientific world and in the popular press. The student who is interested in science can consider either of these as a possible field of endeavor.

Program Evaluation

The emerging hybrid field we will discuss next differs radically from sociobiology and cognitive science in a great many ways but is like them in others. The field is generally known as evaluation research, or program evaluation. Many other types of evaluation—such as cost-effectiveness analysis, systems analysis, and projecting future trends—are akin to the methods we will discuss. The general idea of these fields is to use scientific methods, to the extent that it is possible, to determine what effect a program has actually had or will have. For example, suppose the federal government has financed an educational program such as Head Start with the expectation that it will produce some effect or change on the children who are in it. Since programs like this may have involved billions of dollars over a period of time, it seems reasonable to ask what, if anything, was accomplished.

Evaluation-research methods are also used to determine the effectiveness of pilot programs to see whether they should be continued. For example, a project on the effect of a guaranteed annual income was run on a limited scale with a continuing evaluation. To keep our short discussion manageable we will concentrate on program evaluation as an example of the field.

The basic question in program evaluation is: What are the results of what we did? The answer can be, and most often has been, sought in a nonsystematic and informal way. Our primary interest, however, is in the development and use of systematic scientific evaluation.

As in the case of most scientific developments, the historical roots of evaluation are long. Possibly the first evaluation report occurs in Genesis 1:31, "And God saw everything that He had made, and behold, it was very good."

Some historians of evaluation place the real beginnings of systematic, science-oriented work in the era of the New Deal (1933–1941). During this period the federal government assumed a larger role in a wide variety of fields such as education, relief programs, and regulatory activities. The success of these evaluation efforts is hard to judge in any detail. Shortly after this period (1944–1945) a large-scale effort was made to evaluate the effects of strategic bombing. This early study illustrates a number of factors involved in evaluation research. We shall examine it first, followed by studies of Head Start and the Salk polio vaccine program.

Three Examples

The problem of evaluating the bombing effects was huge, the time short, and available techniques limited. In an attempt to assure an unbiased survey, a committee of distinguished civilians was formed, chaired by Franklin D'Oiler, president of the Prudential Insurance Company of America. At various times individuals such as John K. Galbraith, David Krech, Robert A. Lovett, Paul Nitze, and Adlai Stevenson were involved in planning or executing the survey. More than 1600 civilian and military personnel took part. The idea of organizing a research group of this size is simply mind-boggling.

Conditions for collecting data were less than ideal. Many records had been destroyed either in combat, by design, or through indifference. Others were concealed or in the Russian zone. Hard work, ingenuity, larceny, safe cracking, and in at least one case, the uncharitable might say, kidnapping enabled the group to assemble substantial information.

There were several different reports, varying from a summary report of fewer than 100 pages to a ten-volume final report. Most of those commenting took a positive view of the reports. But there were other comments—for example, "wholly biased. . . . false in great part. . . . childish. . . . unsupportable claims and self-praise." We will return to the problem of differing views later.

Another important evaluation project concerned the Head Start program. Head Start began in late 1964 as a part of President Lyndon B. Johnson's War on Poverty. The general idea was that early-childhood training would alleviate some of the problems of an impoverished environment, enabling children to achieve greater success in school. In June 1968 the Westinghouse Learning Corporation and Ohio University contracted to evaluate the program. Their statement of the question to be answered by the evaluation was "To what extent are the children now in the first, second and third grades who attended Head Start programs different in their intellectual and social/personal development from comparable children who did not attend?" Sounds simple, but it was not.

Before the study was run, several critical decisions had to be made. Among the most critical were in the areas of design and methodology. An ex post facto design was selected. In this design one attempts to find individuals who are similar to the experimental group (in this case Head Start participants) to act as a control group for comparison. This yields quicker results than use of the methodologically more sophisticated randomized longitudinal technique (see the Salk study below). There is a trade-off between time and precision. Another problem concerned whether the measurement should be based on a large number of factors or the two (intellectual and social/personal) that were finally selected.

The report did little to encourage belief in the effectiveness of Head Start: "In sum, the Head Start children cannot be said to be *appreciably* different from their peers in the elementary grades who did not attend Head Start." A controversy followed in Congress, the executive branch, the press, and academic circles. A large portion of the academic controversy revolved around the design, sampling, tests used, and statistical analysis. We can't resolve this dispute, nor is that our primary purpose in considering it. We simply want to show some of the problems involved in evaluation. In the next section we have a general discussion of some of these problems.

One of the best known and most successful evaluation studies was that in the Salk polio vaccine experiment in 1954. Its scope was huge, with 1,800,000 children involved. Several evaluation techniques were used: we will concentrate on the placebo control method. Consider the situation. The vaccine had been substantially successful in preventing polio in nonhumans (primarily rhesus monkeys). Questions remained: How would it affect humans? Were the dosages proper? Would there be any serious side effects?

Half of a large group of volunteers was given the vaccine, the other half a salt water injection (placebo). A double-blind method was used, so that those administering the vaccine did not know which children had gotten saline and which the vaccine. The vaccine group had 16 polio cases per 100,000, while the placebo group had 57 cases per 100,000. Not perfection, but clearly a substantial advantage.

Some General Problems in Evaluation

Several general issues in program evaluation need to be considered. These include issues relating to design, ethical considerations, the reception of the evaluation report, and recommendations by the scientific and nonscientific communities to which the report relates.

One aspect of experimental design concerns control groups. In any evaluation of a program the most direct way to evaluate its success would be to compare the individuals and groups who are in the program to individuals and groups who are similar in every way but are not in the program. Sometimes a control group is impossible, as in the evaluating of the effect of strategic bombing during a war. At other times the decision to evaluate a program is made only after the program has been in effect for a period of time, as with the Head Start program. Sometimes the cost in time and money to establish a control group is prohibitive.

When an equivalent group is not available for comparison, how do scientists tell whether the program has had an effect? What do they compare it with? What do they think the outcome would have been if the program had not been employed? How do they know? What measures do they apply to the program recipients? Generally the researchers would like to find some groups that are similar to the recipients in some ways, as in the Head Start evaluation, but they must always be aware that different contexts can easily lead to different results independent of any specific program.

When one implements a program, it is often not clear exactly what the influence is expected to be. In social programs the people are supposed to be "better off." How do you measure "better off"? When people took the polio vaccine, a good measure was whether or not they contracted polio. When a student has been in the Head Start program, what is a good measure? An IQ test? A reading test? A teacher's evaluation? A parent's feelings about the efficacy of the program? A Children's Self-Concept Index? When do you measure? After one week? After two years? When the participants are 30 years old? And these measures are to be compared with what? The influences may increase or dissipate over time. They may show up in unexpected areas. It's possible that a cure for one disease can cause a second disease after long periods of time; for example, radiation treatments in the '40s and '50s may have caused cancer in the '70s and '80s. We're sure that you can think of examples of programs or procedures that were initiated at one time and had a negative effect at a much later time.

The point to remember is this: there is no universal solution to the design problem; each case must be evaluated on its own terms. Good research in these cases *may not* settle issues; poor research certainly *will not* settle them.

Very often there are important ethical considerations in evaluation research above and beyond the design problems. In the Salk study, if the predictions were correct, children in the control group would have a higher incidence of polio. Is it ethical to withhold a possibly effective vaccine from one group, knowing that this group might suffer more cases of polio? But without the control group the effect of the vaccine may not be clear, particularly over a short term. Also it was possible that the Salk vaccine could have caused severe side effects in the treated group. In that case, instead of an admired figure, Salk would have been considered a monster by many. What are the ethics of experimenting with children? Suppose a cancer vaccine is developed. Would you participate in a mass study? In all such cases, *what are the alternatives?*

One factor is very clear when we compare the bombing survey, the Head Start evaluation, and the Salk study. When there are vested interests in a particular outcome, any other outcome will be challenged. Walter Williams named it "the iron law of absolute evaluation flaws." If you don't like the result, you can always attack "questionable evaluation practices." As you are aware by now, sciences other than evaluation are not immune to controversy, nor should they be. The most important difference is that in most scientific controversies the principal participants are scientists, and

decisions are based, in large part, on scientific grounds. In questions involving public policy newspapers, magazines, television, congressional committees, and other lay groups play a major role. These groups have important contributions to make, because the programs being evaluated affect them directly. But their contributions don't make for tidy discussions.

A Few Concluding Remarks

Earlier we mentioned that sociobiology, cognitive science, and evaluation research are different in a number of ways. First, both sociobiology and cognitive science, although derived from many disciplines, have coherent bodies of potentially measurable data. Evaluation, in contrast, is primarily a set of techniques applied to a great diversity of data. Second, the impact, if any, of sociobiology is likely to be felt much more slowly than that of evaluation, which is here and now. Cognitive science probably fits somewhere in between the other two. Third, evaluation is not likely to make a widespread change in our attitudes toward ourselves as humans. Sociobiology and cognitive science both may have a substantial impact on our view of human nature. Fourth, as indicated earlier, it is not certain that sociobiology or cognitive science will survive. There seems little doubt that evaluation research will continue.

Whatever the fate of these emerging sciences, they will undoubtedly change and evolve. And evolution takes some peculiar and often unexpected paths.

8 Scientists Are People

Previous chapters have presented various aspects of the game of science. But we have not as yet dealt directly with an essential component of the game—the participants. We have talked about scientists from time to time, but our principal concern has been with science itself rather than with the people involved in it. In this chapter we'll take a look at the scientists—their interests, motivations, values, status systems, rules, activities, and education. A large part of the chapter is necessarily based on the authors' personal observation and conjecture.

Motivations and Interests of Scientists

At least two groups who work as scientists can be identified according to their motivational systems: first, those motivated by the intrinsic pleasures of playing the game (the Players); second, those motivated by a desire for recognition and its resulting rewards (the Operators). These two groups are not completely exclusive, but they are worth examining separately. In addition, because individuals are complex, it is doubtful that either of these systems of motivation can account for all the drives that lead any scientist to the intense and prolonged effort necessary to produce a significant contribution.

There are other people trained as scientists who either fail to function as scientists at all or on completion of their formal training no longer attempt to add to the body of scientific knowledge. Their professional motivations are at least partially different from the first two groups. There are quite a few members of these last groups; so, although they contribute very little directly to the development of science, they should be considered.

The Players

Our first concern is that happy breed who find the game endlessly fascinating. For the Players the reason for playing the game is not primarily knowledge or the good of humanity but simply the game itself.

For purposes of organization we can arbitrarily break down the Players' motivations into six topics—which incidentally include most of the attractions of games listed in Chapter 1:

1. curiosity
2. the delights of ambiguity and uncertainty
3. the contest with nature
4. escape from the boredom and crassness of everyday experience
5. aesthetic pleasure
6. the sheer joy that comes from exercising the intellect

1. Curiosity. The first two topics, curiosity and ambiguity, overlap in some ways, but it seems worthwhile to separate them. Curiosity (sometimes equated with the tendency to explore) is a very basic impulse for all higher vertebrates and is also present in many less complex creatures. Such tendencies seem to be primarily innate in humans. Curiosity may sometimes kill cats, but it is more likely to produce new opportunities for learning, location of food, and contacts with suitable mates. There is experimental evidence that more complex species tend to display a higher degree of curiosity than do less complex species. Other evidence, ranging from children rattling and squeezing packages to starving animals exploring a new environment before eating, demonstrates that curiosity is indeed a very strong source of motivation.

Scientists have very strong curiosity impulses, and their curiosity is somewhat different from the curiosity of most others. Two differences are clear: (1) scientists' curiosity about a particular problem may continue through their entire career; and (2) their curiosity in their field is generally impersonal.

Within the scientific community, curiosity plays a particularly large role. For people who have a lively sense of curiosity, science can provide never-ending opportunities to exercise it. The quest for the unknown is as basic to the scientist as it is to the explorer, the philosopher, or the artist.

2. The delights of ambiguity and uncertainty. Scientists not only tolerate ambiguity and uncertainty, they often seek them. Ambiguity generates tension, and for some this tension can be extremely rewarding, even though it continues over a long period. For others, continued ambiguity-produced tension may be shattering. The experience of tension is related to unresolved or incomplete situations. For example, consider a professional football game. What makes a game exciting? Isn't it the tension and excitement due to an uncertain outcome? Why? Which is more interesting and exciting, a contest won or lost in the last 14 seconds or one in which a team wins by a large margin? The uncertain outcome is extremely important. When we watch the TV rerun of a game, our excitement and interest are much less than they are during a live game. The tension resulting from not knowing the outcome plays a large part in our enjoyment. Although science is different in at least some respects from professional football, science also generates the tension of uncertainty. Scientists' informal description of routine projects as "scut work" or "hack work" gives the flavor of their feeling about activities in which the outcome does not contain sufficient uncertainty to evoke tension.

Scientists don't ordinarily seek ambiguity or uncertainty in all phases of their lives. They want their cars to start regularly, their grant requests approved without delay or change, and other facets of their lives to proceed without undue confusion or surprise. In general, their enjoyment of uncertainty lies in the challenge of unresolved problems. Even here the enjoyment of uncertainty and ambiguity has its peculiarities. Scientists are motivated to reduce ambiguity and uncertainty, but they are stuck with the realization that a reduction of uncertainty in one sphere almost always brings about the awareness of new problems with new uncertainty and ambiguity to be resolved. To summarize, the tension resulting from ambiguity and uncertainty is among the important motivational factors in science, and science can be the most uncertain of games.

3. The contest with nature. The old saw of climbing mountains because they are there is illustrative of the fact that a contest with nature can evoke a strenuous response. The principle is very much the same in science, even if the response is ordinarily not so strenuous. Consider the career of a great man and an excellent scientist, Jean Henri Fabre. He spent a great portion of his life studying the behavior of insects. Between the ages of 74 and 92 he worked almost continuously on such observation. This period is of particular interest. What drives a man in his 80s when he is no longer employed in an academic position and has long since achieved fame through his studies? Fabre appears to have taken on understanding the insect world as his own peculiar contest and would not rest so long as he was able to continue functioning. There has to be a very strong quality of attraction that leads a man of 80+ to squat for eight hours watching a solitary spider spin its web. Many amusements lose their savor as we age. In this game lies the possibility of perpetual youthful savor, if not perpetual youth.

Incidentally, Fabre's fascinating works are easily readable by non-scientists. One imagines that, like Bertrand Russell, Fabre felt secure enough in his knowledge and position that he could write clearly, unencumbered by the jargon of his trade.

4. Escape from boredom. How do you escape the sleazy world of TV hucksters, the mundane world of everyday life, and the anxieties generated by personal problems? Some turn to murder mysteries; here, they can substitute their own favorite character for the victim. Other people turn to drink, social events, sex, travel, or Sunday painting. Scientists have a built-in escape in their work. When they are involved in scientific work, other problems vanish. Einstein's address at the Physical Society in Berlin in honor of Max Planck's 60th birthday focused on this particular point: "I believe with Schopenhauer that one of the strongest motives that lead men to art and science is escape from everyday life with its painful crudity and hopeless dreariness." The point couldn't be stated more succinctly.

5. Aesthetic pleasure. Many believe that science and diverse arts such as poetry, music, sculpture, and painting are at opposite ends of a continuum. They are wrong. Like many other human endeavors, science has its aesthetic side. Some theoretical positions or problem solutions are preferred because of their elegance. There are beautiful problems and beautiful solutions. Aesthetic charm in science may rest heavily on lean, simplified lines that are both functional and soaring. Artists and writers differ from scientists in the language they use and in the material they appreciate, but the aesthetic attractions of their game are remarkably like those of the game of science.

The excitement of a strange and lovely object is a form of aesthetic experience. The explorer setting foot on an unexplored planet or the artist stepping back to view a painting must experience some of the same feelings as the scientist who makes an original observation or creates a new idea. It is the experience of oneself as a completely unique individual. Small wonder that Archimedes leaped naked from his bath shouting, "Eureka!" when he discovered his principle. At that moment only he and the idea existed.

6. Joy of exercising the intellect. For athletes, exercise of their prowess can be a thing of joy. They have a pride of accomplishment, and part of their reward is the opportunity to display their capability to admired peers. The same sort of thing appears true of the artist, ballet dancer, or actor. In each case, however, it is the exercise of capabilities for one's own benefit, plus the approval of a few peers, that is important. Scientists don't differ greatly from other professionals. They enjoy the exercise of their intellectual capability in the same way athletes enjoy exercise of their physical capabilities.

There is also a desire for approval by their admired peers. Mere public acclaim may be considered bad taste or even an indication of foolishness, whereas one word of praise from a particular colleague is an event of singular importance. Overall, the motivations of the scientist resemble those of many other professionals; they include the joy of the game itself, together with self-esteem and the approval of valued colleagues.

The Operators

The motivations and goals of the Operators are primarily recognition and its accompanying rewards. (There are good examples of this motivation in fictional characters such as Milo Minderbinder in *Catch-22.*) The Operators have some of the same characteristics as the Players, but they differ greatly in their objectives. This statement is not meant to imply that the Operators contribute nothing to science or that they are unneeded.

The Operators seem, on the average, less gifted intellectually than the Players. In spite of these lesser intellectual gifts, however, the Operators are often quite successful. Operators tend to be very active; in fact, their activity is ordinarily more visible than that of their colleagues. Many Operators get large grants or contracts to do research and capitalize on the contributions of their junior colleagues and students as well as their own. Operators are particularly noticeable in the social realm, serving as working officers of scientific organizations, for which they organize meetings and perform administrative tasks. Many Players do such jobs out of a sense of obligation or sometimes, but rarely, because they enjoy or aspire to these jobs. Acceptance of an administrative position makes serious scientific work difficult or impossible. A few scientists continue to do scientific work in spite of an administrative job, but most cannot. Thus Players are rarely interested in even discussing administrative positions unless they feel an obligation to do so. Richard Feynman, Nobel Laureate in physics and one of the genuine madcap wits of our time, admits he deliberately fouled up on administrative or semi-administrative tasks. After a while the bureaucrats let him alone. That he is capable is indicated by his solo report as a member of the panel investigating the *Challenger* disaster.

Since the Players are under-represented, those who occupy positions as administrators or organization officers often present a distorted view of their discipline to the outside world. These are the scientists who have a greater likelihood of coming into contact with the general public. Many Operators are found in industry, where the pay and other external benefits are higher and the odd wanderings of the self-contained scientist are not acceptable. A combination of rewards is sufficient to attract many Operators to nonacademic pursuits, although many others seek positions of influence or power in academia.

A single distinction between Players and Operators might be that the Players are primarily internally motivated, while the Operators are primarily motivated by external rewards.

The Coaches

There are those who use science as a means toward an end. They enter the world of science as trainees because they think the game is an interesting one to study, but they do not directly play the game themselves. They may prefer to disseminate scientific knowledge to others as a teacher, textbook writer, or reporter. Some people are very skilled at coaching, or science writing, rather than playing the game. These are very important functions, and many Players like to think that one must be a Player to do them well. A diligent Coach may give a potential future Player a vision of what the game is like by delivering the feeling of the game's excitement and explaining the latest versions of some of its fields. It is tough for a non-Player to do this, however, because much of the excitement comes from the dynamics of the game's activity, which in most instances must be experienced to be appreciated. Thus, for a Coach to become successful requires continuing hard work. The final preparation for becoming a Player, however, should be on the field of the game; graduate students almost always must participate as apprentices before they can become full-fledged Players; the final training of successful Players is almost always with scientists who are, themselves, playing the game.

The Bystanders

The fourth group—those trained as scientists who never play the role of scientists (the Bystanders)—is puzzling. What gives a person the drive to complete a long and trying period of graduate study and then terminate direct involvement when the degree has been achieved? The primary factors are probably personal ones, stemming from a lack of independence or self-confidence. Given the uncertainties of the game, a high degree of confidence or even arrogance is required to function outside the protective armor of a graduate school. Consider the situation of young scientists who are not yet established. They have been taught that research, to be of value, must be original. But to be original, they must defy the "knowledge" of their established elders and point to their errors or omissions. This is no position for an intellectual shrinking violet. David had only one giant. The young scientist who advances a new idea arouses an army of giants. In this army are not only foes but one's intellectual forebears as well. The odds in such an encounter are heavily in favor of the giants. Many young scientists opt against originality and, although they remain scientists, do mundane work. Many others trained as scientists find a quiet niche and view the carnage from afar with singular detachment. The latter two alternatives are rational and thoroughly justified. Only madmen would do otherwise. Fortunately we still have our madmen.

There are at least three more possible reasons for individuals' becoming Bystanders. One of these is good old-fashioned laziness. The newly graduated Ph.D. has worked under pressure for several years. There is a strong temptation to relax. Once relaxed, the creature may send down roots and become satisfied and comfortable. Recovery from that state is likely to be delayed indefinitely.

A second reason for terminating serious scientific effort may be a shift in interest to nonscientific pursuits. The shift may be to some activity completely outside the intellectual realm, such as physical culture, Boy Scouting, or the maintenance of the ideal middle-class home. Shifts to activities inside the intellectual community, such as art, philosophy, or history, are relatively easy to understand. Of course, scientists interested in art, exercise, or home life are not necessarily Bystanders. The critical factor is whether they continue to contribute to the body of scientific knowledge.

Third, the reason simply may be boredom. Unfortunately, many who have tried playing any game day in and day out for a long period find that what starts as a delightful pastime may gradually become boring and end as a loathsome task.

Scientific Theorists

Let's now briefly consider a generally honored but sometimes maligned type of scientist, the theorist. Theorists are considered separately, since they are the rarest, most fascinating, and most important of the species *scientificus*. Their motivational system most often is that of the Player, although occasionally it is that of the Operator.

Although theorists are often viewed as cold, rational, deliberate machines, they are generally almost the opposite of this popular picture. They are usually individuals of strong feelings who have the ego of actors and an irrational, almost mystic attachment to particular views of their discipline. The appearance of cool deliberation is their public face, which often represents only their disdain for contact with the spectators.

There are important occupational differences between theorists and other scientists. Theorists set the framework within which others do their research. Those other than the theorists do the important work of filling in details of existing theories. Nontheorists fulfill a relatively safe and useful function. Their work contributes to science but does not threaten the individual scientist unless he or she happens to accumulate evidence contrary to the status quo.

What is the general personality makeup of theorists? Are they normal, neurotic, or even psychotic? They rarely fit the pattern of middle-class normality, and yet they are intensely in touch with their own reality. Perhaps they don't fit any of the usual categories. George Bernard Shaw once said "The reasonable man adapts himself to the world; the unreasonable one persists in trying to adapt the world to himself. Therefore, all progress depends on the unreasonable man." Perhaps his message was to tolerate the dissenters, the faddists, the kooks, and in general those who disagree with what we *know* is right—so long as they don't become too violent. Tolerate them, not out of any sense of humanity but for crass self-interest. A few of them are innovators, and society needs them infinitely more than they need society.

In reviewing all the groups of scientists and science-trained individuals we have encountered, we find a range of individuals spread over the whole spectrum of human behavior but with some important common characteristics. They range from individuals who could easily blend into a meeting of the Rotary Club to others who would find such a meeting more strange and confusing than listening to Little Red Riding Hood discuss comparative anatomy. Scientists are neither supermen nor naive children. They are not foggily absent-minded or unrealistic; rather, many of the things they consider important and real are often quite different from those of the "everyday" world.

The Scientific Subculture

We have spoken of the scientific community as a subculture, and in a very real sense it is. A subculture can be identified as a group that differs in some significant ways from the larger culture and yet has enough in common with it to keep it from being totally distinct. It is perhaps easiest to distinguish subcultures by comparing their value systems and their rules governing behavior. Since many subcultures have a number of characteristics in common, identification can be made best in terms of a total profile rather than a single particular criterion.

There are a number of characteristics of modern science that have had a very decided influence on establishing and enforcing the prevailing value system within the scientific community. None of these is uniquely characteristic of the scientific community. The scientific subculture is characterized by (1) a severe and formal selective process, (2) disciplines within the subculture involving small numbers of active participants, (3) far-reaching communication and mobility, (4) an internal system for establishing relative status, and (5) a set of formal and informal rules that govern status and support the value system. Although the fourth and fifth points are included in this list, they will be discussed separately later. Characteristics of status, values, and rules are closely intertwined, and their separate classification is largely a matter of convenience.

The severe and formal selective process that precedes one's acceptance as a scientist limits the number of possible participants in the scientific community. It also makes for acceptance of and adherence to the tribal mores. Scientists go through a period of five to nine years of indoctrination in their particular discipline. The portion of this time spent in graduate study is devoted to close contact with individuals who have undergone a similar indoctrination. In addition to the formal requirements, graduate programs usually force students into close social contact with others who are in similar circumstances. Their world comes to revolve around their particular discipline. There are few experiences, other than taking monastic orders or going to certain professional schools, that give this degree of orientation to a discipline. It would be surprising indeed if people did not respond to the values and rules of a system that plays this large a role in their life. One of the more obvious effects of such total submersion into the system is that the associated values assume a particularly large role in developing the individual's view of life. It is not only indoctrination that leads to the acceptance of the mores of science. The nature of the scientific enterprise demands it. If one does not accept them, one simply does not become a successful practicing scientist. Formal and informal educational forces that mold the scientist will be discussed in the section called "Stumbling into Science."

A second characteristic of the scientific community is that the number of individuals working in a particular problem area is relatively small. It is typical that these individuals are acquainted with the work of most others with similar interests and very often establish personal contact or correspondence. One of the principal functions of conventions and professional meetings is to maintain these personal contacts (and usually much more information is exchanged and more interesting ideas explored in the bars, in hotel rooms, and at parties than in the formal convention meetings). Another product of the personal relations within disciplines is that theoretical issues very often have strong personal implications. Again we emphasize that science is not as impersonal as often pictured. Attachments and disputes can take on intensely emotional tones and may be conducted with the dignity and logic of a fishmongers' argument.

The third characteristic involves the wide-ranging communication and mobility of scientists. As mentioned before, science is not a completely solitary game, and scientists tend to flock together like geese during migration. Very often these concentrations ignore national boundaries, and the communication and contact lead to establishment and maintenance of an international system of values and rules. In the early 1900s, when Germany was the leading scientific center of the world, many scientists were attracted there to study and often remained. From the end of World War II until about 1970 the United States, because of its lead in many areas (due in large measure to money and safety), attracted researchers from all over the world. Recent cuts in scientific funds have greatly curtailed this exchange. Certain institutions also attract many researchers. There are only a few institutions that are important in areas such as nuclear physics, astronomy, and oceanography. Even in disciplines where expensive installations are not important, the presence of one or two important theorists or researchers makes an institution very attractive to the young scientist. There are other aspects to the mobility of scientists. Within rather wide limits, scientists can move freely from one institution to another without seriously disrupting their work or being forced to re-establish their reputation. In contrast to most fields of endeavor, science is not primarily dependent on local evaluation or evaluation by those unacquainted with the practitioner's craft. Due to ease of communication and personal contact, scientists' judges are their peers wherever they may be. It is a wise dean or department chairman who realizes that his or her own judgment of a scientist is of slight importance unless it coincides with that of the scientist's peers.

Values

Earlier we discussed some of the ways in which scientists acquire their particular value system, without specifying what values they acquired. There are a number of highly specific customs peculiar to any specific discipline. For example, in physics it is generally accepted that observations are to be put in mathematical form at some point. On the other hand, such a rigid ground rule would seem strange or perhaps stupid to the anthropologist. In each case the scientist is acting in accord with the tribal customs, or values, of the discipline. Fascinating as the variations among disciplines might be, we will consider only those values that cut across a wide variety of disciplines.

One of the values is individuality. The scientist is expected to reflect his or her own judgment and analysis within the field. This doesn't mean that he or she must hold views different from those of others; they don't have to be unique in every aspect, but they should be the scientist's own views, not mere echoes.

Another characteristic valued within many sciences is the ability to consider the data and interpretations without personal involvement. The principle is that the scientist should be able to back off and look at the data and interpretations as though they had come floating in a bottle from the sea. There are some scientists, personified by theorists, who definitely do not fit all parts of this description, but it remains a value even among them, despite the violations.

One of the most important values is the dedication of the individual to the discipline. The system is directed toward fostering dedication; laziness or lack of interest meets with all the sympathy it would have received from the early Puritans.

A question that is often raised in relation to values is why more scientists do not have, and implement, ideals that would lead to resolution of some immediate practical problems. The problem is not one of ideals but of behavior. As indicated earlier, many scientists are interested in knowledge as an end in itself. In other words, they are trained and interested in problem solution, not solution application. Many feel that solutions for problems such as war, poverty, and education already exist. The problems of solution application are seen as political and social. The critical difficulty is that scientists attacking an applied problem directly are required to move outside their paradigm. This is like asking an accountant to become a salesperson, an economist a merchant, or an architect a brick mason—which can happen but rarely does and even more rarely is successful.

We could consider a number of other items in this section on values—for example, honesty in research is probably the single most important value in all sciences—but, because of a high degree of overlap, we shall discuss these other items in the later sections on status and rules. Both status and rules reflect underlying values, and adherence to values is rewarded through achievement of status.

Status

The scientific community, as is true of most subcultures, has its own internal system for establishing status. Its status hierarchy derives from its value system; however, we are not primarily concerned here with the source of status hierarchy. Status in different communities can be based on a wide range of criteria. Among so-called "society," ancestry gives status within the group. Among some South African groups, obesity gives status. A skinny leader is absolutely unthinkable. Among businessmen, wealth (or the appearance of wealth) is a primary status symbol. Among Plains Indians, the number of horses stolen was a source of distinction. There are even status hierarchies among other primate groups. For example, among mandrills, male status is related to the brilliant blue of their faces and the equally brilliant red of their behinds. The status of an individual may be due to a characteristic that is arbitrary (ancestry), achieved (horses stolen), accidental (tendency to obesity), utilitarian (wealth accumulated), or hereditary (mandrill's coloration).

In science the status hierarchy is a combination of a number of factors, including the following: (1) the discipline of the scientist: this is largely arbitrary but is related to the stage of development of the discipline; for example, physics rates higher than sociology; (2) the role played: a theoretician generally rates higher than other scientists; (3) originality and influence on others: influence on others ordinarily takes precedence; this order is reversed only when the original work has sufficient impact to influence others so that it becomes a combination of originality and influence; (4) the institution with which the scientist is associated: the hierarchy may be difficult to define, but most scientists can identify it. One measure that correlates with, but does not define, the hierarchy is the amount of research funds available to the institution.

Rules for Behavior

In addition to the positive aspects of achieving status, there are some informal prohibitive rules proscribing certain behaviors; these are derived from the value systems. Violation of these rules ordinarily results in various degrees of loss of status or rejection by the scientific community. These rules are not exclusive to the scientific community, but the great emphasis placed on them requires separate discussion.

About 1980, there was a rash of articles, reports, and books reporting on, discussing, and suggesting remedies for dishonesty in science. Over the last couple of years either scientists have become more honest or cautious, or the fad has died down.

There are two cardinal sins in science—lying and stealing. Of these, lying is the more serious. In the case of science, lying does not refer to claims such as "I sat up with a sick friend." Rather, it is the knowing misrepresentation of data. The reason that this is so serious is that it distorts the research process. Work may be done on false premises, time and money may be expended uselessly, and trust may be shaken.

Stealing consists of taking credit for the work of another. Although stealing is serious in an ethical sense, the source of data is not as critical as its existence and reliability. We certainly don't approve of either lying or stealing, but we do recognize the difference.

Some of the most important scientists in history, such as Ptolemy, Galileo, Newton, Mendel, and Dalton, have been accused of misrepresenting their data. We really don't know whether these accusations are true. We are much more concerned with the present; in the past the ground rules were not as clear. In any case these people must have done some honest days' work, since we still utilize their ideas and findings.

After reviewing a good deal of literature, we are not sure just how much dishonesty there is in science today. Our personal observations lead us to believe that it is not extensive. Others, such as Broad and Wade (in bibliography) conclude that it is widespread. In an appendix to their book they list cases. For the twenty-year period prior to the publication of their book they list twenty cases. They included two students, one historian, one parapsychologist, and three cases from outside the United States. Whether this is an indication of "widespread" is a matter of judgment.

We have no idea whether scientists are more dishonest or less so than others in their nonscientific activities. But within science there are some constraints. First, the penalties are rather severe. Lying will in almost all cases result in permanent banishment from the scientific community. Second, to have any effect, you must publish your work, thus leaving a continuing record. Third, if your work is judged to be important you can count on its being repeated. If the results don't stand up, usually it will be assumed that you made an honest mistake, but others may then regard your work with increased caution.

Any dishonesty in science is a matter for serious concern, but then so is stupidity. We believe that there is more of the latter than the former.

Choice of a Research Area

The choice of a research area may seem an unlikely subject for a chapter on scientists as people; please be patient. Research covers a large number of popular topics and a larger number of topics with only a few interested participants. Why do particular areas of research exist? Is it because the natural phenomena simply cry for investigation? This may be part of the answer, but at best it is only a part. Researchers often create an area by showing that a seemingly limited topic contains a variety of interesting phenomena. There are topics that have been abandoned, others that attract only one or two researchers, others that have yet to be opened, and still others where the researchers swarm like gnats. Why is a research area attractive? A particular area may appear to offer the possibility of posing and answering interesting questions. In addition, a particular topic may fascinate some individuals. A complete explanation for the attraction to different research areas would be impossible. The question is like asking why a man finds a particular woman attractive; the possible explanations are unlimited, but they are not likely to be conclusive.

The vast number of possible research areas that have received little or no attention may suffer for one of three reasons. First, we may not know that the subject matter exists. There is no cure for this lack of awareness other than tripping over it in our wandering, as Anton van Leeuwenhoek did in the discovery of bacteria. Second, we may know of the problem but lack the technical capability to investigate it. For example, scientists assumed that the moon has a back side, but prior to the development of rocketry the possibilities of direct observation didn't exist. Third, although we may know of a problem and have the techniques to investigate it, few become interested in it. Some problems are studied by only one or possibly a handful of investigators. How do we account for these hermits of science? If one is the only investigator in a field, the problem of competition does not arise; for some, safety is important. It is also possible that Players develop a feeling of proprietorship in a particular area when they are the primary investigators. In such a case they may even resent the intrusion of other investigators and may discourage their entry.

Our problem is further complicated by fads and fashions. There are sudden fads in science as well as in the everyday world. A new empirical "discovery" may be reported. For a time the particular topic may get "hot" and investigators swarm in. There are a number of investigators who are always entwined with the latest fashion. Many other investigators find a topic that suits them and ignore the fashion changes. A new topic generates concepts that may prove fruitful or may become a will-o-the-wisp. By the same token a steady, self-chosen course may only lead to a deeper rut. The game is not without a large element of chance.

Most scientists could have chosen any of a number of different disciplines. The election of a particular discipline from a fairly large set is likely to be fortuitous. As with marriage, proximity at the right moment, some undefined attraction, convenience, or the absence of competition can all be cited as reasons. In science we might find that a book, an article, a lecture, or simply the path of least resistance lies behind the decision to work in a particular area. Again parallel to marriage, there is a difference between the initial decision and a continuing relationship. The continuing relationship can be based on genuine delight in the relationship itself. There is also the possibility that the relationship is controlled by inertia, fear of new experiences, or indecision.

Stumbling into Science

This section is called "Stumbling into Science" because, like choosing a research area, becoming a scientist has a large element of chance. Despite this element, however, we can look at some identifiable characteristics of people who are likely to become successful scientists. There are exceptions to each of the statements that will be made, but these statements do seem to be true of a large group of successful scientists.

The prime requirements for the successful scientist are curiosity, intelligence, an interest in understanding concepts, and the capability and motivation to work long hours. Alas, such attributes, when found in the young, are often threatening to their teachers and classmates. In addition to the prime requirements, the prospective scientist is likely to be independent and somewhat aloof. These are not the general characteristics that lead to nomination as most popular student. Surprisingly, in spite of the apparent handicaps, prospective scientists tend to be rather well adjusted socially.

There are no formal rules for recognizing or measuring any of the above requirements except intelligence. This can be measured relatively reliably by scholastic aptitude or intelligence tests. There has recently been a controversy over how to interpret intelligence tests. However, for those who have no obvious cultural deprivation—and perhaps even for those who do—the tests do predict rather well if they are given and interpreted by competent testers. For most doctoral programs in science, students tend to be in the upper 10% on college aptitude and achievement tests.

Formal Education

Today the process of becoming a scientist is much more formally defined than in earlier times, although there are differences from institution to institution and among disciplines. Some of the general statements that follow do not fit individual cases, but they will give some idea of what is involved.

One or two centuries ago the scientist was often a gentleman amateur, and even those who had formal training had received most or all of it as apprentices. The gentleman amateur now either is extinct or has assumed the role of spectator, and apprenticeship alone has become a very difficult route to science.

Today, in order to have a successful career in science, it helps to start young. The most successful future scientists are apt to learn a lot of math and read widely when young. They usually get good grades in school and go to top-rated colleges and universities. They show diligence and persistence throughout school and score high on standardized tests, particularly on the quantitative parts.

The formal aspect of science education ordinarily begins at the undergraduate level at a college or university. Future scientists need a broad educational background. They should have a good orientation in such fields as the arts, history, philosophy (particularly 20th-century philosophy), literature, and mathematics. They should develop a familiarity with computer science. If the student has a knack for languages, they may be helpful; otherwise, they should be avoided. Two years of required language training as undergraduates will be largely useless if students stop there. If, however, they are fortunate enough to get an extensive background in Russian, Japanese, or Chinese, they will find it very useful. German and French are often recommended, but these languages do not now have as much utility as some others. After being on the skids for various reasons, science in France and Germany is coming back, but it has not reached its former level of dominance, and there are many translations and translators around. That is not as much the case for the other recommended languages.

A future scientist must get into a graduate school, the higher quality the better; this requires good grades in college. A minimum average of B is required to be accepted in most graduate schools, and it is often necessary to have some A's in solid courses such as math to accompany a B average. A failing grade in any hard-core courses may be fatal, whatever one's average.

The undergraduate scientist should not attempt to become a specialist; that will come soon enough. A scientist should be *educated,* not just trained. Unfortunately, not all graduate schools admit students according to this principle. The plea here is for a liberal education at the undergraduate level, not in the classic mold but attuned to the broad reaches of current knowledge and thought. This sort of program should precede specialization so that one has at least the capacity to recognize the existence of different paradigms and the relations between various fields. The alternative to a liberal education is often like a typical business-administration program, in which the student gets approximately twice as many specialized hours as does the physics major. Perhaps business is more complex than physics; or perhaps the object of a business school is to produce the equivalent of a trained seal. Among their other excellent qualities, trained seals perform on command, present a uniform, neat appearance, and slide smoothly through the water, creating no waves.

Probably the most important step in the formal education of scientists is their graduate program. They should shop as carefully as possible for an institution and should try to get information from recent catalogs and faculty members, since departments change very rapidly. They should pick a department with a number of active young researchers. There are departments composed primarily of grand old names; if the history of science is one's interest, this type of faculty has many advantages; otherwise it should be avoided. By the end of their first year of graduate work, students should have a pretty fair idea of where their specific interests lie. They are also likely to become apprentices and junior colleagues of some faculty member. Perhaps the instructor will present an outlook that inspires the students to make science a lifetime hobby as well as work.

Most of the factual material a student is given through undergraduate and graduate education will rapidly pass out of date. A Ph.D. who drops out of a field for about five years or less usually requires substantial retraining before being able to return to basic research. (Of course there are relatively static areas where this statement can be challenged.) Several years ago, one of the authors changed his research area. Even after a relatively short period he no longer feels competent to teach graduate courses, conduct research, or even discuss his old area intelligently. The moral should be clear: the principal objective of an education is to prepare the student for continuing self-education. Incidentally, the problem of self-education is closely tied to research. Active researchers are forced to keep reasonably current in their field. Researchers who are not familiar with what is going on in their field will have their omission pointed out by an editor or colleague. Editors, in particular, can be rather nasty.

Some Final Thoughts on Education

The fortunate student will have instructors who make their subject live. It is very tempting to follow the lilting tune of these Pied Pipers. And probably no harm will be done, since their interest is likely to be contagious.

For those students who have the capabilities of becoming scientists, many specific disciplines could provide a source of unending entertainment. If possible, the decision should be delayed for a time. Once they have arrived at a decision, students should head straight for the laboratory and library. They should wash test tubes, clean cages, bring in coffee, build equipment, prepare specimens, sweep, mop, compute results, and read,

read, read; in short, they should do any and all the jobs they can, whether paid or unpaid. This may not be the perfect test of their interest in the particular discipline, but it will give them first-hand knowledge. A caution: many universities and researchers do not encourage undergraduates to get into the research laboratory. The student should be patient and willing; free help will not go unrewarded forever.

Reading cannot be overemphasized. Students should not limit themselves to their chosen area. A wide range of reading will not make them experts in every field, and it may not even be of any direct help in their chosen field. But it is a good safeguard against becoming simply a well-trained technician.

If you expect to be a professional scientist, don't end your involvement when you walk out of the classroom or lab. An important part of your education will be informal discussions in hallways, at parties, over lunch, in coffee shops, and at bars. These discussions give fledgling scientists a chance to test their ideas, safely challenge authority, learn how others are thinking about problems, and learn about areas other than their own.

Employment of Scientists

The previous section, about how and why people go into science, was based on our own view of the field. How scientists are employed can be tied more closely to data. When this section was written for the first edition, nearly all groups and individuals (the authors included) felt that the problem was to fill the demand for scientists. There are several areas in which the need for scientists still far exceeds the supply, such as in biotechnology and computer science. For several years now, however, the problem has been more one of finding the demand. Overall, the number of science students has been decreasing; but the number of jobs available in many areas of science is still insufficient. Some earlier projections indicated that overall science jobs and science graduates would be in balance by 1985. This did not occur. In 1983, about 88% of doctoral scientists were employed in jobs related to science. However, in 1973, 93% were employed in such jobs. The fact that 75% of federal research funds now go to the military indicates a further erosion in the kinds of jobs available. The entire situation is quite complex; reams have been written bewailing, denying, and explaining. We shall try to hit a few high spots and hope interested readers will look into the problem in more detail.

One of the most troublesome and difficult problems encountered in attempting to make plans for future scientific needs is that both institutions and individuals must always project several years in advance of the actual events. For example, the average time between entering college and completing a Ph.D. degree is about nine and a half years. Students have some flexibility over the first few years, but as time passes their options become more limited. The lag time for institutions is also very long, which affects decisions to start or stop programs or to change their direction. From personal experience, we can state that it requires about ten years from the time a new Ph.D. program is conceived until graduation of the first students. The fact that one field, nuclear physics, changed from a shortage of scientists to an oversupply in about three years is enough to make all involved wary. A number of other factors contribute to make the guessing game resemble blindman's buff. Future economic conditions, governmental attitudes and commitments, student preferences, and public attitudes, in addition to changes within science or changes directly affecting science (new discoveries, new fields, new technologies), all complicate the situation.

In spite of these difficulties, the situation must be dealt with. A few suggestions come immediately to mind. We need some up-to-date and reliable figures concerning just where we stand. Many institutions are in extreme financial straits; canceling expensive Ph.D. (or other) programs may be both rational and timely. And governmental agencies, particularly the National Science Foundation (NSF), need longer-range planning authority. Two other items are worth considering. Most scientists still get jobs—not always the jobs they want and not always in their specialty, but actual unemployment is lower among scientists than among the remainder of the population. As of this writing, the unemployment rate for scientists with doctorates is about 2.6%. While some fields have been particularly hard hit, others have been affected only slightly or not at all. Second, there is nevertheless a serious danger that a failure to continue recruiting young scientists in the academic area may decrease the probability of innovation. No matter how loaded with honors the elder statesman of science may be, it is still the brash young upstart who is most likely to produce something new.

Should the foregoing discourage the prospective scientist? Not necessarily. Competition will be tougher, and the marginal student will find both graduate school and the job market very difficult. The dedicated (or driven) student of high capability should continue to find satisfactory chances to become a part of the scientific enterprise. Incidentally, the NSF's *Review of Data on Science Resources* indicates that the percentage of graduate students enrolled in nonscience areas is increasing—in contrast to the situation in science—and that this is a long-term trend that began about 1963. The meaning of the trend is not clear.

The NSF periodically surveys the various areas of scientific activity in the United States. The following data are taken primarily from various NSF publications; these sources are not always in agreement, and they don't always agree with other sources. We have done our best but don't guarantee anything. In each case we have used what we considered the best and most recent data. The most recent information available at this writing usually pertained to 1982, 1983, or 1985.

As you will note, the definition of *scientist* used in this section differs markedly from the one we have used up to now, because what (or who) counts as a scientist here is determined by the U.S. government. Government reports usually consider scientists and engineers together. They also include as scientists people with only bachelor's or master's degrees, whereas most of these people would be considered by us to be technicians.

Much of the 1985 *Science Indicators* and the 1982 *Survey of Scientists and Engineers* is based on questionnaires submitted to scientists and others. This method creates problems. Despite the glowing description of scientists you've been reading here, they too have their vanities. You may recall the statement that basic science is higher in the status hierarchy than applied science. If you're a scientist and you work in both applied and basic science, are you more likely to describe your "primary work activity" as basic or applied? Two guesses. Further, the salaries mentioned vary greatly by institutions, by geographical areas, and by specialties within disciplines. Here we will ignore such important considerations in favor of general tendencies.

Of approximately 3,465,000 "scientists and engineers" surveyed by the NSF, about 11% held doctoral degrees, 34% master's degrees, and 55% bachelor's degrees. Whether those with degrees below the master's level should be considered scientists or technicians is an open question. Note that these figures also include engineers.

Of the group of doctoral scientists and engineers surveyed by the NSF in 1985, the majority were employed in educational institutions; about one-quarter were in business and industry, and the rest were scattered over government (about 10%) and a variety of other areas.

There were more life scientists (25%) than any other category in the 1985 survey. Anthropologists, linguists, sociologists, and statisticians each constituted 1% or less of the sample. The 1985 *Science Indicators* also shows that basic health research in medical schools is passing rapidly into the hands of Ph.D.s in the life sciences. Basic science departments show increases in the number of Ph.D.s and decreases in the number of MDs since 1972; however, the number of MDs in clinical departments has increased over 50%. One outcome of such shifts is that the percentage of principal investigators on National Institutes of Health grants who are MDs has dropped from nearly 60% to about 40%.

While we are on the subject of money, let's briefly consider doctoral scientists' income. They rarely get rich, but on the whole they are reasonably comfortable. Again the situation is complex, and we can look at only a selective summary of approximate figures. In academic institutions, the 1985 survey indicates, the median salary was about $40,100 per academic year over all areas of science for those with doctorates. As might be expected, pay was better in industry (overall median, $44,500). Civilian scientists in the federal government were paid about the same as those in industry.

Activities of Scientists

The NSF surveys provide data for a rough estimate of the primary work activities of doctoral scientists. About one-third of the scientists report that they are engaged in research and development; another third are in teaching. About 6.5% are now in management or administration. The latter is up about 50% since 1972. An interesting and significant trend, but we are not sure what it means. The research-and-development item needs somewhat closer examination. The best estimates we can make are that about one-third of those in that area are in basic research, while the remainder work in applied research or development. Industry devotes only about 3% of its research-and-development funds to basic research.

Some mention should be made of the scientist-turned-administrator. Combining active research with administration is a rare and difficult enterprise. The problem of administration in science is quite serious. Administrators are not likely to be among the current leaders in research, and only the current leaders are likely to be capable of fully understanding the research problems. One solution used by some effective administrators is to use active researchers as advisors. Another possibility is to return administrators to other jobs on some periodic basis. This is particularly urgent in academic settings where we are plagued by those who have become professional administrators. After a while they tend to view universities in the same light as they would shoe factories.

Women in Science

In spite of a strong cultural bias that has existed in the past and still exists to the present day, women have made and continue to make major contributions to science. One of the most important scientists of this century is Marie Sklodowska Curie, the first woman to win a Nobel Prize, and also the first person to win two Nobel Prizes.[1] Her fame is well deserved, and not merely because she discovered two previously unknown elements (radium and polonium). By realizing that radiation is an atomic phenomenon—that the atom, in other words, is not really *atomos,* or "uncuttable"—she opened the Atomic Age.

Marie Curie's daughter Irene also won a Nobel Prize, in chemistry (1935), shared with her husband Frédéric Joliot-Curie. Irene's contribution pales only beside her mother's, for she and her husband discovered artificial or induced radioactivity, making possible the radioactive tracers and radioisotopes that have become so important both to basic knowledge and to medicine.

[1]She shared her first prize, in physics (1903), with her husband Pierre, and with Henri Becquerel, who had discovered radioactivity. Her second prize, in chemistry (1911), was awarded to her alone. With a modesty too seldom heard on the Nobel podium, she insisted that Pierre deserved joint credit with her.

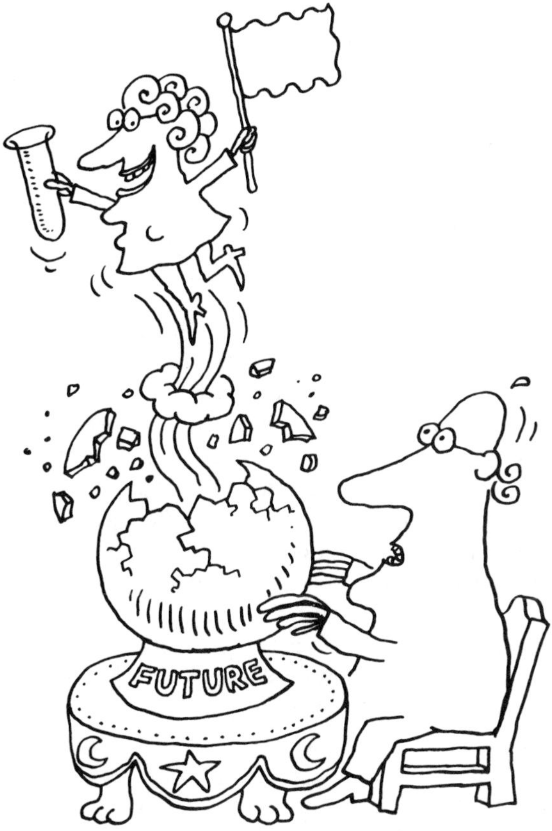

Besides Marie and Irene Curie, six other women have won Nobel Prizes in science: Gerty Cori in 1947; Maria Goeppert-Mayer in 1963; Dorothy Hodgkin in 1964; Rosalyn Yalow in 1977; Barbara McClintock in 1983; and Rita Levi-Montaleini in 1986.

Gerty Cori received the prize in physiology or medicine for elucidating the process of sugar metabolism in humans. Specifically, during years of joint research, she and her husband Carl had traced out the so-called Cori cycle, in which glycogen in the muscles is metabolized to lactic acid, which is carried by the bloodstream to the liver, where it becomes liver glycogen. This in turn is released into the blood stream as glucose, which is eventually transformed back into glycogen in the muscles. Along the way they had discovered, among other things, phosphorylase, an enzyme vital to this metabolic process. Gerty Cori also did important independent work not included in the Nobel citation. She determined the structure of the glycogen molecule and differentiated four types of glycogen-storage disease.

Maria Goeppert-Mayer won a Nobel Prize in theoretical physics, honored with Hans Jensen for her discoveries in nuclear shell structure, the so-called Mayer-Jensen theory. Her particular flash of genius was to see the relationship between the measured spin of an atomic particle and its orbit around the nucleus. Goeppert-Mayer and Jensen did not work together, but made their discoveries independently, Jensen's paper arriving at the office of *Physical Review* only two days later than hers.

British chemist and crystallographer Dorothy Crowfoot Hodgkin won the Nobel Prize in chemistry for "determining the structure of biochemically important substances by X-ray methods." These intricate and difficult methods have been compared to deducing the shape of a jungle gym by observing the shadows of its bars. Two of the undeniably important substances whose crystalline structure she determined were penicillin and vitamin B-12. To read the early history of penicillin is to be impressed with the extremely elusive and difficult nature of the penicillin molecule and thus with the magnitude of Hodgkin's achievement—which the Nobel Committee called "a magnificent start to a new era of crystallography." Her determination of the vitamin B-12 structure was considered "the crowning triumph of X-ray crystallographic analysis," not only because the results are so important in chemistry and biology, but because the structure is so complex. Since knowing the structure of a molecule enables scientists to synthesize it, the medical importance of her work is also obvious. Since her Nobel award, Dr. Hodgkin has been working on the insulin molecule.

American physicist Rosalyn Yalow won the Nobel Prize in physiology or medicine for developing a technique called radioimmunoassay. Her technique, developed jointly with a physician, the late Dr. Solomon Berson, measures minute amounts of substances such as hormones in the blood. The technique is so sensitive that it could measure a teaspoon of, say, insulin in a lake 62 miles long, 62 miles wide, and 30 feet deep. With radioimmunoassay—which uses the radioisotopes made possible by Irene and Frédéric Joliot-Curie—doctors can at last monitor insulin blood levels accurately, for the better understanding and treatment of diabetes.

Barbara McClintock was given the Nobel Prize in physiology or medicine in 1983. Although it was intended specifically for her discovery of mobile genetic elements, the award was also given in recognition of a long series of brilliant and creative contributions to cytology and genetics. She carefully specified the morphology of maize chromosomes. She and her student Harriet Creighton firmly established the physical basis of genes in the chromosomal bodies in the nucleus of cells. She reported that pieces of chromosomes can become separated from the rest and, by forming "ring chromosomes," might become inactive. She identified organizational, inhibitory, and building functions for genes. She reported how mutations may be due to control-process variations within the cell and to direct structural changes on the affected gene. She discovered two new kinds of genetic elements that control some of these functions.

Relatively early in McClintock's career, but after her potential was clear, a dean was quoted as saying: "Barbara McClintock has proved to be a troublemaker and ______ hopes an offer may come her way so that she can have her career elsewhere than at Missouri." A horrible thought intrudes—can it be that administrators are not always the best judges?

The woman who was most recently named a Nobel laureate is Rita Levi-Montaleini, an Italian-American developmental biologist. She and Stanley Cohen won the Nobel Prize in physiology or medicine in 1986. They won for the discovery of chemicals that stimulate the growth of cells. In particular, Levi-Montaleini discovered what is now called nerve-growth factor. This is a protein that, when secreted by another cell, stimulates a nerve cell to grow toward it. These growth factors seem to play an important role in such diverse biological phenomena as embryological development, the growth of cancer cells, and the growth of different organs such as the brain and the skin.

The scientific contributions of women are not limited to those who have won Nobel Prizes; several women have done much of the work that led to Nobel Prizes for others. Rosalind Franklin did almost all the spectrographic analyses and interpretations that allowed Watson, Crick, and Wilkins to win the prize in 1962 for discovering the structure of DNA. She very likely would have shared in that award had her untimely death not occurred. Lise Meitner did much of the conceptual and empirical work for the discovery of uranium fission that led to Otto Hahn's winning of the prize. Meitner and another colleague shared in a Fermi Award with Hahn for that discovery. British graduate student Jocelyn Bell discovered the pulsar that led Antony Hewish to the Nobel prize in Physics in 1974; and Dr. Chien Shiung Wu devised a complex experiment that first disproved (1956) the concept of conservation of parity in the universe, but she did not share in the 1957 Nobel Prize in physics. This concept of nonparity has figured in another Nobel Prize (1980), and an article on the 1980 awards in the December 1980 issue of a popular science magazine called *Discover* mentions the 1956 experiment but fails to mention Dr. Wu.

The list, unfortunately, could go on. Nor is the Nobel the only science prize whose winners' list contains fewer women than it might. Candace Pert discovered endorphins, one of the most important developments in brain chemistry of the past several decades. But the Lasker Prize awarded for the discovery in 1978 went instead to the director of her lab, Dr. Solomon Snyder. Dr. Pert, now at the National Institute of Medicine, continues to be a creative scientist, showing that endorphins are involved not only in the brain's natural handling of pain and in opiate addiction but in the brain's processing of sensory information as well.

Fortunately, the list of women who have contributed new ideas to science could also go on. In the brief scope of this book we will content ourselves with mentioning just two more: a nuclear physicist and an astronomer.

Dr. Leona Libby has been at the forefront of nuclear-energy technology since the atom was split. She was a prominent member of the team that built the first nuclear reactor, the first and second Argonne reactors, the

Oak Ridge reactor, and the three Hanford reactors for producing plutonium. Among her firsts, some alone and some in collaboration, are the discovery of very cold neutrons, the building of the first thermal column, and the building of the first rotating neutron spectrometer. In addition, of great value to other scientific disciplines, Dr. Libby discovered that historical climates can be measured from isotope ratios in tree rings.

One of the first things to impress itself on the student of astronomy is the awesome distances involved—and the difficulty and crucial importance of knowing these distances as accurately as possible. Henrietta S. Leavitt, who worked at the Harvard Observatory in the late 19th and early 20th century, was the first to realize the relationship between the period of variation in a variable star and its brightness. This so-called period-luminosity law formed the basis for Harlow Shapley's later work on astronomical distances, and thus for much of 20th-century astronomy.

Some changes are under way in the education and employment of women in science. For example, at the doctoral level, the percentage of women receiving science degrees rose from about 7% of the total in 1965 to 25% in 1983. The distribution is not even over fields, however. In 1984 life sciences, including psychology, had about one-half women graduates, but engineering had only about one-tenth. In both fields the percentage of increase from 1965 was substantial (life sciences, 575%; engineering, 2800%, partly because of the low initial rates). The trend is clear: the future holds a larger representation of women in science. Employment has also changed. Educational institutions have led the way, with close to one-fifth of science and engineering faculties being women; business and industry are lowest with less than one-tenth. The rate of growth of advanced degrees and employment of women in science is greater than men in all categories. Before you cheer yourself hoarse you might note that women's salaries are only about three-quarters that of men's. A number of factors, such as seniority, may account for at least part of this discrepancy.

Women are still most heavily represented in lower-echelon positions in almost all fields. As their presence becomes more commonplace though, they will undoubtedly occupy more positions at all levels, and their contributions will be more duly noted.

The Daily Grind

Research scientists as people are much like the rest of us. They eat, cannot find a pair of socks or stockings, curse the idiots in traffic, exchange meaningless polite greetings, entertain sexual fantasies, change diapers, remark on the undeniable cretinism of bureaucracy, attack sexism, and in general behave like tame creatures from our society. Then there are activities characteristic of their occupation. They think about problems, design experiments, talk to research assistants and technicians, look at equip-

ment, check apparatus, cope with malfunctioning equipment, analyze data, puzzle over confusing or conflicting data, discuss problems with colleagues and assistants, hold meetings or seminars, worry about research budgets, and read, think, and write. These last three activities are their most important functions; if scientists are fortunate, these activities occupy a substantial portion of their time. If they are academic scientists, somewhere in the schedule they meet their classes and try to communicate some of their knowledge and enthusiasm to their students. If we are all fortunate, they may contribute something to the acquisition and dissemination of scientific knowledge.

Of course, not all is perfect. Science, like any other profession, requires a great deal of plain hard work. The ten-hour work day is more common than that of eight hours. And not every day is glamorous or exciting. Sad to report, frustration is a more common experience than exhilaration. Scientists may become bored, confused, tired, fed up, and angry. They may conclude that the whole game is pointless. Fortunately, these states are usually transitory: on balance, the attractions outweigh the blemishes.

9 Science in the World

Numerous discussions in this book have indicated that science has a very strong impact on the everyday life of nonscientists as well as scientists. The effects are seen in the way people view themselves and their universe (the intellectual aspect) as well as in more obvious technological progress. Even the attitudes and views of the scientifically naive are influenced by science, although these people are generally unaware of the source of this influence. Galileo gives us an excellent example of both the intellectual and technological influence of science even in his day. Galileo's astronomy was of the utmost importance in changing the view of humanity's place in the universe and in helping to decide the struggle between science and authority as bases of knowledge. On the technical side, Galileo's improvement of the telescope and his studies of ballistics had an undeniable practical impact.

Science and Social Change

Over the past 150 years it has become increasingly apparent that people have the capability to design or change both their material and social environments. With the rapid development of science and technology, the number and range of possible alternative decisions they must consider have increased fantastically. The capability for change is still not absolute, nor will it necessarily lead to a better system. (Read Aldous Huxley's *Brave New World* and George Orwell's *1984*.) However, people now have to face the painful fact that they can and must make decisions regarding their material and social environments, and often they must do so without any precedent for the decision. A decision is always made, since failure to act is also a decision.

Scientific Advancement

Let's look at an example of a probable future development to see how this development would lead to problems and changes in the intellectual, social, and technological aspects of society.

Take the subject of aging. Let's assume that aging can be controlled so that an individual can have a useful life span of about 200 years. Incidentally, such a development seems likely, provided the major powers forgo the ecstasy of mutual extermination. Let's also assume that this increase in life expectancy could be accomplished within the next 50 years. (We will ignore the logical problem of how we would know we had extended life by 100+ years before at least 100 years had passed.) It is estimated that the total scientific and technological development necessary to achieve this goal would not cost more than a yearly expenditure equal to 5% of the present yearly U.S. military budget. Recognizing the relative attractiveness of life destruction versus life preservation, as well as the immediate self-interest of some groups, we don't expect any great diversion of funds. Even if military expenditures are decreased, a cry of "we can't afford it" will probably block or delay any large-scale effort. But ignoring this pessimism (or realism) concerning funds, assume that at some date we will control aging. What would be some of the intellectual and technological results in society?

From a technological viewpoint, consider some of the economic consequences. Any impact on economic theory per se will be ignored here, since our purpose is one of illustration. One very obvious economic consequence of controlling aging pertains to the training of people in the work force. At present typical Americans complete most or all of their formal education prior to full-time entry into the work force. But even today it is necessary to send engineers and other technical personnel back for refresher courses on some occasions. New developments must be communicated or else technical personnel rapidly become outmoded. The same thing is true to an even larger extent among scientists. For example, the social sciences recently developed use of computers, statistical techniques, and mathematical models, leaving many older scientists bewildered, since they don't have the background to understand these new developments.

Consider the plight of those who completed their education about

1880. Assuming they were alive today, could they be qualified to perform any complex task in today's economy through self-education? Undoubtedly, a few individuals could; in most cases they could not. Such problems seem certain to become much more evident and more serious when aging is controlled. New concepts of education will be necessary. Instead of an early period of education followed by a few "refresher" periods, a regular routine of return to education may be required. We could follow the educational problem further, particularly the question of how to keep the educators themselves current, but even this brief a discussion should suggest a host of related problems.

It has been said of politicians that "few die and none resign." Industry has been somewhat more successful in maintaining a turnover in top management, primarily through the use of arbitrary retirement ages. One important part of our present problem is that long-term incumbents (political, industrial, or otherwise) tend to use old solutions for new problems. In addition, long occupancy of an office establishes a feeling of ownership together with the power to defend that ownership. Is it desirable to have the president of a corporation or university remain in office for 100 years? We think not. The effect of very extended periods of tenure in advanced positions would also be felt in ranks of "young" aspiring workers. How many of these would have the patience to wait 20, 30, 40, or 50 years for promotion?

The control of aging will surely require some basic changes in our economic organization. The examples given should indicate the breadth and complexity of these problems.

Our present attitudes toward something as basic as human life are related to the period of growth and to life expectancy. Although it seems certain that many of these attitudes would change if aging were controlled, it is not clear exactly what forms such changes might take. We can look at a few problems, however, to see attitudes that would be influenced by the control of aging.

Since people ordinarily don't live more than about 75 years, extension of this period to about 200 years would have to be controlled through some sort of specialized treatment. The first obvious question is: Should everyone get the treatment? Or more bluntly, who lives and who dies? The first impulse is to answer that everyone has the right to live. But what of individuals who suffered permanent and serious damage, deformity, or other disability through illness or accident? Should they be preserved if there is not serious hope of recovery? What if they are in intense pain? Or have become a human vegetable? These may seem like extreme cases, but they do illustrate some of the difficulties. Other cases that arouse mixed feelings are individuals who pose a clear threat to society. Do we prolong the life of a person imprisoned for a violent crime? The result might be a choice between an endless sentence straight out of Dante or the guarantee of release at some point without regard to the consequences for others.

These problems exist even today; they would become much more serious if life were extended.

In the early stages, and perhaps for a long period during a program of age control, it is quite unlikely that facilities or personnel would be adequate to extend the program to all the people in the world. Would life itself become a new tool in nationalistic competition? If facilities were not sufficient to take care of entire populations, how would decisions be made? How do you decide which life is worth preserving? A few years ago a somewhat similar problem faced a group charged with deciding who would be scheduled to use a limited number of artificial kidneys. The choice was literally one between life and death. The agony involved in these decisions was obvious. Multiply this by millions, and consider the impact.

Admittedly, the first examples have been chosen to indicate some sharply defined effects of our hypothetical development. Other effects might not be so dramatic, but they could conceivably have as great a total influence on our attitudes. How would people value life if they had the prospect of 150 or more years of useful activity? Would they cling to life with greater tenacity, or would boredom dilute its savor?

Inevitably, population control would also become a problem even more pressing than it is today. How would we ration procreation? Would it be left to individual judgment? Would our attitudes toward new additions to the population remain unchanged? It doesn't seem likely. These questions don't have simple answers.

This speculation about age control has considered only a few facets of technological, social, and intellectual changes as examples. The possibilities for other changes are almost limitless. It seems highly probable that our entire society would be permanently altered, for better or worse. We must face the fact that, with or without major events such as the control of aging, our society is rapidly and constantly changing. The rate of change is accelerating, and a substantial portion of the changes are due to science or to science-related activities. Unfortunately, despite rapid change, our society suffers from lack of application of our knowledge and capacities, particularly in the social sciences. This failure to apply our knowledge is the principal topic of the next section.

Cultural Lag

As an example of the lag or failure to apply our knowledge derived from science, consider the problem of nuclear war. In the early 1980s, several groups of scientists considered the long-term biological effects of a major nuclear war. There were two earlier studies, one sponsored by the National Academy of Sciences, and the other by the U.S. Congressional Office of Technology Assessment. One of these groups of scientists, referred to as TTAPS (Turco, Toon, Ackerman, Pollack, and Sagan), caused quite a stir.

The TTAPS report went beyond the known effects of nuclear war, such as radiation blast and firestorms, or even radiation sickness. Their conclusion was that beyond these annoyances there were likely to be long-term effects of a serious nature. The explosions would raise large amounts of dust, smoke, and debris. These would form a cover obscuring the sun. If this did in fact occur the results would be catastrophic.

In April 1983, about 100 scientists from the United States and several other countries met in Cambridge, Massachusetts, to review the findings of the TTAPS group. They represented a wide variety of scientific fields, including physics, biology, meteorology, and others. After this meeting, an important book, *The Cold and the Dark,* which considered the consequences of nuclear war, was written. The following is a brief excerpt:

> The group generally agreed with the conclusions of the [TTAPS] report as to the potential for substantial reductions in the amount of solar light reaching the Earth's surface and for severe climatological changes, although suggesting minor adjustments. In addition to climatological effects of freezing temperatures and virtual darkness, the physical science group discussed stresses such as radiation exposure and fallout, exposure to ultraviolet radiation from sunlight owing to a depletion of the ozone layer, and impacts from toxic gases released by combustion of synthetic materials.
>
> . . . [A] group of biological scientists, plus ten of the physical scientists . . . considered the prolonged darkness and severe climatic changes and their effects on phytoplankton and zooplankton and other plant and animal life and on agriculture. . . . Effects on plant and animal life from long-term exposure to ionizing radiation and ultraviolet light were considered. Other discussions focused on large scale interruptions in the normal services of natural ecosystems, which are crucial to the support of human life and society, including production of food for humans as well as for domestic livestock and wild animals; climate and weather; disposal of wastes and recycling of nutrients; soil preservation; and control of crop pests. . . . [T]hey concluded that they could not rule out the possibility that *the long-term biological effects of nuclear war could cause the extermination of humankind and most of the planet's wildlife species.* [emphasis added]

Since the TTAPS report there have been several other studies in several countries. Questions have been raised as to the magnitude of the effects. Generally speaking, however, the other reports, both in the U.S. and abroad (including the Soviet Union), have supported the idea that the results of a major nuclear war would be long-term and devastating. Interestingly, the U.S. Department of Defense has supported some studies. While they don't exactly contradict the idea of nuclear winter they often conclude we "just don't have enough information." Somehow this fails to surprise us.

The recent nuclear incident at Chernobyl (April 1986) was insignificant in comparison to a major nuclear war. It was about the equivalent of the single bomb dropped on Hiroshima. Today the United States and the Soviet Union combined have stockpiles in excess of 20,000 bombs, most of which are 50 times as powerful as the Hiroshima bomb.

The consequences of Chernobyl are still being counted, but it is clear that the effects covered thousands of square miles, killed many people, affected crops and livestock, and forced the removal of about 135,000 people. The Laplanders of northern Scandinavia (about 1000 miles away) may disappear as an independent culture because of the devastation caused by the Chernobyl incident to the reindeer population and the rest of its food supply. That's about as close to an experimental demonstration as we care to observe.

Many other examples of lags between the accumulation of knowledge and its acceptance and implementation could be cited. A few will be mentioned briefly.

Air and water pollution: We don't know all the answers, but we can make substantial improvements now. After many years of getting worse, air pollution and water pollution in the United States seem to be lessening in severity. Los Angeles has fewer critical smog days, and the air over many other cities is likewise improving. Certain lakes, streams, and rivers—including Lake Erie, which had been declared dead—have more fish than they had previously. Things are not perfect. With the increase in the use of coal we have to worry about "acid rain," which has destroyed the life in many lakes and streams. Economic difficulties have also led to a clamor for lowering the levels of pollution control. But we have begun to apply scientific knowledge that has been available for several decades.

Health: The accomplishments of medical science have been remarkable. Health-care delivery has not matched these accomplishments, though it seems to be getting better. The World Health Organization recently announced the elimination of smallpox, but many other curable communicable diseases still kill millions every year. One global measure of the application of medical and social technology is the infant mortality rate. Even with improvements in medicine, the rate in the United States remained stagnant or increased until the late 1960s, when it started heading downward. Now it is lower than it has ever been, about 10.5 infant deaths per 1000 live births. However, it is still higher than the rate of about 25 other nations around the world. Most of the nations in Western Europe, including all of the Scandinavian countries, the United Kingdom, France, Spain, and West Germany, have rates of less than 10 infant deaths per 1000. Other nations including Canada, Australia, Hong Kong, Jamaica, and Israel all have better rates than the United States. Japan, a nation of 122 million people, has a rate of 6.0, almost 43% better than that of the United States. We cannot use economics as an explanation of the relatively poor U.S. rate, as it spends more on medicine than any other nation in the world. In the poorer nations of the world up to 20% of the infants die each year, a rate that is over 1900% greater than that in the United States. There are obvious cultural lags when we consider these numbers, both in the United States and around the world.

There are in the United States other cultural curiosities, such as the reluctance to accept the metric system, the response to criminal or other antisocial behavior, the antiquated educational systems, extermination of wildlife, and the resistance to fluoridation of water.

But *why* are there serious delays between the accumulation of scientific knowledge and its acceptance or implementation? Some of the general sources of resistance to the acceptance and use of scientific findings can be listed:

1. difficulty of communication
2. inertia
3. confusion due to disagreement or apparent disagreement among scientists
4. influence of special-interest groups
5. fear of change

First, there is the complex problem of communication. How many people are likely to read, understand, or even know of a study such as that on the nuclear winter? A thorough knowledge of a scientific controversy such as this is even rarer and, in addition, requires an understanding of meteorology, biology, and physics that only a handful possesses. Because of such problems, the acceptance of scientific statements becomes a matter of faith for most and usually requires a generation or two to be effective, since faith changes slowly. Today most people accept on faith the idea of the earth traveling around the sun, but they are no more able to defend such a belief than their ancestors were able to defend the belief that the earth was the center of the universe.

There are at least two other aspects of the communication problem. Today there is a very serious lag in publication of new material. Often "new" research is two or three years old before it is printed in journals. The newest textbook material is generally three to five years old. Also, communication between basic and applied scientists is difficult, since these groups read different journals, attend different meetings, and have different interests and backgrounds. Even when they attempt to communicate with each other they often fail, since they see the same phenomena through different paradigms.

Second, there is a problem of inertia. This is so common to any enterprise that lengthy explanation seems pointless. We need merely keep in mind that any change in the pattern of human activity, thought processes, or institutions requires extra effort. The direction of a rut may be irresistible.

Third, new knowledge may be surrounded by an aura of controversy. The disagreements may be quite basic, as illustrated by Louis Agassiz's attack on Darwin's theory of evolution. On the other hand, the areas of disagreement may mask more general agreement on important issues. For example, a later debate over evolution involves the gradualists and the followers of punctuated equilibrium. The gradualists see very slow changes across large numbers of a species. Punctuated equilibrium predicts relatively sudden changes (in geologic terms) within relatively small isolated groups. Exchanges between these groups are often rather heated. Yet they agree on a commitment to the idea of evolution as well as the kinds of data that should be collected, the animal groups that are related, the importance of tracing evolutionary changes, the mechanisms for these changes, genetic theory, the general principles of classification, and most other things. Application is often delayed when there is debate about the underlying principles, whether the debate is relevant to the application or not.

Fourth, all of us have the capacity to find a "good" reason for our resistance to change. These reasons often reflect only a special interest. It would have been surprising indeed had the tobacco industry agreed with the Surgeon General's report that smoking is a danger to health. Equally unsurprising was Southern politicians' resistance to the Supreme Court's school-integration decision of 1954, because, in their words, it was based on sociology and not law. In these cases and many others, the special interest is rather clear but in some cases the special interest is less obvious. Economic, social, religious, educational, and other groups and individuals also have vested interests in particular aspects of the current scene. One would have to be a super Pollyanna or simpleminded version of Candide's philosopher friend Pangloss to expect that individuals or groups would readily give up their pet ideas, regardless of the contrary evidence or reasons. Whether people can learn to be more flexible or rational remains to be seen.

A fifth obstacle that delays acceptance or utilization of scientific findings is the fear of change or new situations. There is very clear evidence that unknown or changed situations evoke both curiosity and fear. For present purposes we will ignore curiosity. The evidence of fear covers a wide range of animals, from rats to humans. The fear produced by a new or changed situation may be as clear as that of a primitive tribe faced with its first airplane, or it may be the less-defined fear that questioning or even discussing a favorite belief in terms other than clichés will bring down an entire belief system. It is difficult to define clearly the formless fears of unknown things. Perhaps a reasonable analogy is a child's fear of the dark. The fear may be expressed as "something might get me," or "there is a tiger in the room." Whatever the stated reason, it probably bears no relation to past experience. Fear of change seems to have the same irrational but real character. No matter how we state our objections to a new development, we should remain aware that part of our response is probably based on a very primitive form of fear.

While considering some of the obstacles to change, let's not assume that change for change's sake is necessarily good; rather, we must be aware that irrational resistance to change can prove fatal to an industry, government, economy, or any institution—even science.

Failure or delay in coping with change will become an increasingly serious problem in a rapidly changing world. Flexibility and the capacity to accept rational change may very well be the margin between progress and disaster. This is not the place to resolve the problem. It is simply suggested as a lifetime research project for a few psychologists and sociologists.

Science and Ethics

People do many things; they behave, they are aware of both behavior and potential behavior, they have feelings about both behavior and nonbehavioral events, and they make decisions about these feelings and behaviors. One kind of decision people make about events or behaviors is an evaluative one. People evaluate behavior and events according to whether they are pleasant or unpleasant, whether they are liked or disliked, and also whether they approve of them or not. Any decision based on such an evaluation is considered a *value judgment.* If the evaluation includes whether a behavior is good or bad, it is considered an *ethical judgment.* Many behaviors, of course, are not given an ethical judgment. There are other ways to define values and ethics, but these will do for our purposes. If the ethical judgments made by members of a society about an individual's behavior are that it is good, the individual will usually be considered a moral or ethical person. The exception is that many behaviors are considered so bad that only one or a few instances of such a behavior will identify a person as unethical or immoral. High "goodness ratings" of behaviors are called "ethical standards."

Ethical standards are usually accompanied by some means of reward or punishment, which may range from social approval or disapproval to the use of physical force to enforce the standard. Since behaviors are accompanied by feelings, values always underlie ethical standards, but values can exist without any ethical implications. For example, eating a very rare steak may be highly valued by one individual but rejected by another in favor of a very well-done steak. Acceptance or rejection would not ordinarily have any ethical implications. On the other hand, the knowing sale of tainted steaks would undoubtedly be considered unethical by both individuals. Still others might also condemn the eating of any beef on ethical grounds. For better or worse there are few, if any, human activities that would not be considered by some to have ethical implications. The individual determines whether a particular behavior or event is related to values and ethical standards.

Society tends to look at the relationship between science and human ethics in an inconsistent manner. On the one hand there is a great deal of fear that science falsely defines "good behavior"; on the other hand, some nonscientists expect science and the scientific methods to be the ultimate arbiter of ethics. Some people look to science for their ethical principles as others look to religion.

The relation between science and ethics is quite complex. Most of our attitudes, including those related to ethics and morality, have been affected by scientific thinking and discovery. Applied scientists have made many discoveries that are directly relevant to human behavior, but scientific concepts are not themselves behaviors and therefore are ethically neutral.

The ethical position of scientists as people is quite independent of their scientific competence. One can consider the behavior of scientists according to two sets of criteria: first, their behaviors as citizens, evaluated by society; second, their behaviors as scientists, evaluated by the scientific community. If they collect data in an unbiased manner, and if they can understand and interpret scientific achievements, they can be good scientists no matter what their ethical position on other matters.

There are ethical considerations in a scientist's work, but ethical standards are not discovered by playing the game of science. Scientists may be able to say with scientific validity where a particular activity will lead, but they cannot use scientific principles to decide whether either the activity or its result is good. However, science can play a direct role in ethical determinations in at least two ways. First, given an ethical principle, scientists can determine some external conditions that are associated with behavior conforming to the principle, or conditions associated with behavior violating the principle. For example, let us assume that murder is bad. It is within the province of the social scientist to discover the conditions ordinarily associated with an increase in the number of murders. It has been shown that the uncontrolled availability of firearms is one such condition; therefore we can conclude that the uncontrolled availability of firearms is associated with consequences we have labeled "bad."

Second, scientists can establish what the consequences of certain behaviors actually are, independently of ethical considerations; on this basis, people can make their own ethical judgments. For example, children who are reared with limited stimulation tend to be permanently retarded in their physical, social, and intellectual development. By using all the information available, we may or may not decide that it is good to provide children with various levels of stimulation during their early years.

Some ethical judgments are made by specific individuals, some are made by particular groups, and some seem to be made by society at large. Many of these decisions are very difficult to make, and it takes a great deal of wisdom to make good decisions. The basis of this wisdom seems to be twofold: an appreciation of civilization and culture achieved through study of history, literature, art, and the other humanities on the one hand, and some scientific knowledge of the events associated with the problem on the other. This basis suggests that the humanist and the scientist should have some knowledge of each other's fields and that the lay person should be aware of both.

Although people use science to make ethical judgments, they do not make these judgments according to the principles of science. This statement needs some examples for clarification. There is scientific evidence that people who smoke cigarettes have a shorter life expectancy than those who do not. A smoker has a greater tendency to contract any number of diseases, including gastric ulcers, heart disease, and lung cancer. In addition, many people enjoy smoking. Some have reported that cigarette smoking may relieve tension. Also smoking is used by some people as a means for establishing social status and acceptance, and even as a symbol of virility and masculinity. Nowhere in this description is there a statement about whether it is good or bad to smoke. It is simply a description of the actual consequences of smoking. It is for scientists and others to decide on some nonscientific basis whether these consequences are good or bad.

For years, scientists have been doing research in attempts to discover and understand the cause or causes of cancer. Many of these studies require procedures using live animals. These animals may be injected with cancer cells as part of the experiment and then sacrificed in order that their tissues may be studied. Scientists can't use scientific methods to help them decide whether it is ethical to give these animals cancer. Rather, they use ethical standards that they have obtained elsewhere. They can use scientific goals as one of the bases for making their ethical judgments, but the decision about whether it is ethical to sacrifice animals is argued on other than scientific grounds. The use of animals is sometimes defended on the ground that new information or insights about the disease may be attained. This argument asserts that knowledge of disease is good, so it's ethical to acquire this knowledge; but scientists can't show scientifically that such knowledge is good. It is sometimes argued that this knowledge may lead to techniques to alleviate the disease in humans. This argument implies that disease is bad, but scientists can't demonstrate that disease is bad, either. The argument that sacrificing animals now to save the lives of humans later is often made. This argument suggests that human life is worth more than that of other animals, but again that assertion cannot be demonstrated. Sometimes the justification given for experimenting with animals is simply to gain more knowledge of the natural world. This justification and all the others stated here are generally accepted by our culture, but in some cases the real motive might simply be that some scientists enjoy watching animals suffer.

There is nothing in science stating that it is good to attempt to save human lives. Saving human lives seems to be a generally held value in most cultures of the world, but it is not in any sense scientifically derived. Nor is there any scientific derivation stating that it is not good to sacrifice animals. Some scientists feel no need to justify the sacrifice of animals under any conditions. Most hunters see no need to justify killing or crippling animals. Standards of scientists, like the standards of others, vary.

Consider another example. Scientists have discovered some of the causes of certain birth defects such as mongolism, and they can determine long before birth, with little or no risk to either fetus or mother, whether a fetus has such defects. But science does not say whether it is ethical to induce an abortion. The methods of science cannot specify any conditions under which abortion is ethical. Scientific methods can only lead us to understand the actual consequences of different actions. Look at the other side of this ethical dilemma. Is it ethical for a physician who knows that a fetus is deformed not to recommend abortion? The question is outside the realm of science, whichever way it is asked.

Should a scientist sacrifice a dog to find out what the effect of a drug is? Should a scientist build a hydrogen bomb? Should a scientist develop a chemical that in minute quantities can kill millions of people? Should a scientist develop a contraceptive pill? Science cannot answer these questions. A scientific conclusion is "If such and such is done, then this will follow." Ethical criteria from other sources are needed to decide what ought to be done.

In summary, if we ask the question "Do scientists playing the game of science have anything to say about ethics?" the answer is a resounding yes. Applied science, in particular, is important, because the applied scientist attempts to devise ways to achieve the ethical goals desired by individuals. Ethical goals have to be established in nonscientific ways, but, once they are established, the scientist has the best opportunity to find out how to reach them. The proof that science is independent of any particular ethical system can be seen in the fact that a scientist can use science in an attempt to reach an amazing array of goals: to save a life or to take a life; to make people fearful or to make them unafraid; to build a better bomb or a better food, a better mousetrap or a better mouse. Once the goal is established, scientific methods provide the best means for reaching it. But none of these goals can be defined as good or bad by scientific methods.

The Relation between Science and Technology

Science and technology are mutually supportive. Science benefits from technological advances, and much modern technology rests on scientific discovery. Until relatively recently, much of technology developed independently of scientific advances, although much of science depended on the then current technology. Today, however, new technology usually depends on scientific discovery, and this dependency will undoubtedly increase. In the past, technology was relatively simple and utilized readily available materials. This is unlikely to be the case in the future.

To illustrate the mutual relationship between science and technology, we might consider the microscope. An early step in the developmental sequence was the invention or discovery of a method of making glass. This event was preceded by the ability to control fire, as well as other necessary antecedents. Early records of the Egyptians indicate that they knew how to make glass, and this knowledge probably came from Asia. But the origins of this major technological development are buried in prehistory.

There is a report that a convex lens made of rock crystal was found among the ruins of the palace of Nimrod (about 1800 B.C.). Certainly, for many centuries prior to scientific research, men had known of the magnifying powers of natural or accidental lenses. A significant step in this history is a book of optics attributed to Ptolemy. Ptolemy did make significant contributions to science. His book on optics (written in the second century A.D.) laid the scientific groundwork for later work (about A.D. 1000) by Ibn-al-Haytham, who advanced the science of optics through extensive research. Roger Bacon (1270+) was familiar with Ibn-al-Haytham's work and described a telescope, although there is no record that Bacon built one. The first modern convex lens (a simple magnifying glass) was made by grinding (a technological advance) sometime in the late 1200s. A major breakthrough in microscopes occurred about 1590 when Hans and Zacharias Janssen, Dutch spectacle makers, made the first compound microscope. In 1611 Kepler outlined the construction of an improved compound microscope, based on optical theory, and a model was built in 1628 by Christoph Scheiner. This microscope established the rudimentary pattern for the modern optical microscope. Note that up to this time there was an intermingling of scientific and technological steps: the techniques of glass and lens making were necessary to the science of optics, and the scientific principles of optics were equally important to the design of an advanced microscope.

We could move from the development of the microscope to a discussion of its use, but it is sufficient to say that the microscope plays a very significant role as a tool of both science and technology. There are various discrepancies in treatments of the history of lenses and microscopes, but these discrepancies shouldn't disturb us, since we are interested in the principle rather than the accuracy of this historical illustration. And the principle is not affected by minor variations in the sequence of events. If you find this acceptance of confusion about history strange, read H. L. Mencken's "Hymn to the Truth," which casts a beam of revealing darkness on the history of the bathtub.

To gain some appreciation of the technological products that rest on a scientific base, one has only to look around. For example, whatever the limitations of fluorescent light, it is vastly superior to kerosene lamps. Consider a few of the basic scientific discoveries necessary to produce this illumination. Again, the rude beginnings go back probably beyond recorded history. There is evidence that about 500 B.C. Thales, a Greek philosopher and scientist, rubbed amber with a cloth and observed that the amber then attracted lightweight objects, such as strands of feathers. Doubtless others, in idle moments, had seen the attraction of amber for other objects, but we have no record of their observations; possibly they only shrugged and went on to more important matters.

The curious behavior of amber was known for about 2200 years before William Gilbert (an English physician) made an important extension to our knowledge of this peculiar phenomenon. Gilbert found that other substances (such as sulphur and glass) have the same properties as amber. Strangely enough, lodestone seemed to have some of the same characteristics as amber, except that lodestone and amber attracted different bodies. It became clear that scientists could not limit their studies to the attractive qualities of amber. By 1646 Sir Thomas Browne (another English physician) contributed the name "electricity" to the phenomena being observed. (Browne also believed in witches and assisted in their examination, but he rejected lycanthropy.) Charles Du Fay (a chemist who was also superintendent of gardens to the King of France) studied the attraction and repulsion of charged objects; by 1733 he came up with the idea that there are two kinds of electricity. At this time (around the 1740s), as Thomas Kuhn phrases it, "There were almost as many views about the nature of electricity as there were important experimenters." The first and possibly greatest American scientist, Benjamin Franklin, proposed a concept that organized a large portion of the known data and gave direction to much later research. Franklin concluded that there is only one "electric fluid." This fluid is found in all bodies; those bodies having an excess quantity are positively charged, while those with smaller quantities are negatively charged. The concept was crude and did not account for all the information available at the time, but despite its deficiencies Franklin's way of organizing the data was accepted as being superior to others. Franklin's successors developed his ideas about electrical phenomena in a number of ways, but his concept was a critical beginning. Franklin has been best known popularly for his later experiments and for his explanation of the similarities between static electricity and lightning (particularly the kite experiment, 1752). As ingenious as this later work was, it was not nearly so important as his earlier conceptualization. As pointed out before, concepts (or theories) are more important than isolated facts.

In passing we should glance at the work of the Italians Luigi Galvani (a professor of anatomy) and Alessandro Volta (physics), if for no other reason than the fact that their work originated in the twitching of frogs' legs. The anatomist concluded that frogs' legs contain electricity that is released when they touch metal. The physicist concluded and demonstrated that the chemical action of moisture and different metals such as iron and copper produces electricity. Note that the difference in analyses of the events is related to the difference in paradigms of the investigators.

In 1820 Hans Christian Oersted (a Danish scientist) showed that an electric current has a magnetic effect. By 1822 André Marie Ampère (a French physicist) worked out the laws on which the present-day concept of electricity is based.

Michael Faraday, the English chemist and physicist, gives a fitting climax to our consideration of the history of electricity. Faraday believed that if electricity could produce magnetism, then magnetism could probably produce electricity. In 1831 he found that relative movement of a magnet and wire loop leads to induction of an electric current. A lesser-known American scientist, Joseph Henry, also discovered this principle in 1831. The principle discovered by Faraday and Henry is the basis of the construction of electric generators and motors. Why is it that this scientific history omits a discussion of Thomas Edison? He can be placed high among the developers of the *uses* of electricity, but his scientific contributions to the *concepts* of electricity are few.

Before drawing any conclusions about the scientific work discussed, let's consider for a moment the technological fallout. Could a modern technological society exist without electricity? It's difficult to see how. Certainly we could have developed alternatives for some things, such as fluorescent lights, but alternatives to radios, telephones, computers, television, radar, most internal-combustion engines, and many chemical processes would be out of the question.

From this abbreviated history of electricity we can extract a few points, some of which have probably already occurred to you.

1. Investigation of electricity was truly an international and interdisciplinary endeavor. Greeks, Italians, French, Americans, English, Germans, Danes, and Dutch all made significant contributions, as did philosophers, amateur scientists, chemists, anatomists, physicists, physicians, and generalists. As has been mentioned before, science is an open community, with membership dependent on contributions, not on politics, geography, or professional titles.

2. The pace of the history picks up over time. The history is long, but events come closer and closer together as we approach the modern era. This increasing intensity is in part due to the interdependence of science. Each new discovery or concept suggests additional possibilities or relates prior findings.

3. Curiosity motivated most of the researchers. What distant horizon hid the "practical" applications? Could Thales, Gilbert, Franklin, Galvani, Oersted, or even Faraday have had immediate "practical" applications as their goals? This motivation does not seem even remotely possible. These men had found their game. Our present-day technological fallout is an unsought and unforeseen by-product of their efforts. If their efforts had been directed toward immediate application, where would we be? Not only would we have failed to produce basic scientific knowledge of electricity, but we could not have achieved our present technological development. Had the Roman engineer/administrator outlook dominated our culture, we would doubtless have achieved infinite refinements of existing knowledge and techniques, together with some fortuitous gains. Just how we could have achieved new insights and changes based on abstract principles is impossible to conceive.

4. To restate an earlier point, technology and science have strong mutual influences, even though they have quite different objectives.

The Role of Basic Science

A few years ago in an article in *Science,* Julius Comroe and Robert Dripps considered the "Scientific Basis for the Support of Biomedical Science." They selected cardiovascular and pulmonary diseases for their study, since these were their areas of competence. They also had wide acquaintance with others in the field who could serve as sources of advice.

Without going into the rather complex but apparently objective research (they screened over 4000 scientific articles and analyzed 529 that they and 140 consultants considered essential to the field), we can proceed to their conclusions:

> (i) Of 529 key articles, 41 percent of all work judged to be essential for later clinical advance was not clinically oriented at the time it was done; the scientists responsible for these key articles sought knowledge for the sake of knowledge. (ii) Of the 529 articles, 61.7 percent described basic research; . . . 21.2 percent reported other types of research; 15.3 percent were concerned with development of new apparatus, techniques, operations, or procedures; and 1.8 percent were review articles or reported synthesis of the data of others. . . . Basic research, as we have defined it, pays off in terms of key discoveries almost twice as handsomely as other types of research and development combined.

Amen.

Lest the skeptical conclude that we have no interest or experience in applied research, we might point out that for the past sixteen years one of the authors has been involved in research on the effects of crowding. Although a large part of the motivation for undertaking this research was interest in the basic processes, the results achieved that noblest of all goals—relevance! Briefly, it was found that larger numbers of people in a housing unit resulted in higher rates of noncommunicable illnesses. Some types of crowding are associated with both natural cardiovascular and cancer deaths. Some are also related to suicides. These findings were used (even before they were published) in a New York court case to prevent increasing the number of prisoners in housing units. At present, several billions of dollars worth of new prison construction and remodeling is being heavily influenced by that research. This application was not even considered until the research was well along. It was then literally thrust upon the researcher with both some good (new sources of data) and some very annoying consequences.

The same author worked for three years on explosive decompression (the sudden reduction of pressure) that would occur if you had a blowout in a space capsule). The results, although different from common presuppositions and the Air Force manuals, clearly pointed to the needed emergency equipment and procedures.

We remain convinced of the vital role of basic research. We also have observed that it is easier for individuals trained in basic science to work on applied problems than it is for those trained in applied approaches to shift over to basic research.

Jerome B. Wiesner, special assistant for science and technology to President Kennedy, displayed courage, wit, and keen insight in his struggle to explain the nature of basic research and its possible relation to technology. Had you been in his position, using the example of electricity, how would you have explained to budget administrators or members of Congress your reasons for financial support of Gilbert's amber rubbing, Galvani's frog-leg preparation, or Franklin's kite flying? One's imagination fails at this point. Fortunately Wiesner's did not. His summary of the problem rings with the clarity of a rock-crystal goblet.

> The research scientist is primarily motivated by an urge to explore and understand, but society supports fundamental research primarily because experience has demonstrated how essential such work is for continued progress in technology. Halt the flow of new research and the possible scope of technical developments will soon be limited and ultimately reduced to nothing. Incidentally, scientific knowledge need not be exploited immediately once it becomes available. It exists for all to use forever.
>
> Thus, acquiring scientific knowledge is a form of capital investment. Unlike most other capital investments, it does not become obsolete; nor can it be used up. Technological developments are also a form of capital investment, though somewhat less enduring. To be sure, a more efficient process will also yield its benefits endlessly, but usually technological developments tend to become obsolete as better methods, devices, and processes emerge.

Financial Support of Science

We now come to the grubby but vital topic of money. First, we must remember that science, on any significant scale, exists only in economies where there is a surplus of goods and services. A prosperous economy forms a necessary base for scientific work. Tragically, the economies of the "underdeveloped" countries, where the need is overwhelming, have not supported—nor have they been capable of supporting—any substantial scientific and technical training or research. Read the speech of P. M. S. Blackett, president of the Royal Society (cited in the Bibliography), in which the existing great disparity between the "have" and "have-not" economies is considered. His primary theme is that the gap is not closing; on the contrary, it is steadily growing wider.

United States and Its Rivals

In earlier editions we indicated that the United States was spending more on research and development than any other country. This is probably no longer true.[1] The Soviet Union is clearly spending a much larger *percentage* of its gross national product in this area. Because of problems such as exchange rates, even experts are not sure just how the actual amounts compare, but if there is any difference it is very small. It is clear that the Soviet Union has more scientists and engineers than the United States. Whatever the absolute levels, the trends are unmistakable. In 1969 the United States spent about 2.9% of its gross national product on research and development, and the Soviet Union only slightly less. By 1985 the United States had dropped to about 2.5%, and the Soviet Union had increased to about 3.7%. Thus the Russians were almost 50% higher in this category. In 1969 the Soviet Union had about 10% fewer scientists and engineers in research and development than the United States. By 1983 they had over 40% more. Figures 9–1 and 9–2 show the dismal picture. The United States still has one advantage, though even that is narrowing over time: whatever the level of mindless meddling and obstruction by the U.S. bureaucracy, it runs a very distant second to the Soviet level. (See the books by Zhores Medvedev or Cracraft's *The Soviet Union Today* listed in the Bibliography.)

Japan and West Germany are also making vast strides in catching up with the United States in their spending on research and development and in the number of scientists in that area. One of their substantial advantages is that each spends relatively little of its government research-and-development funds on defense research (12% and 25%, respectively), while the United States spends about 75%. Defense research is usually a dead end as far as its contribution to science is concerned. Each spends over half of its research-and-development funds on basic research, whereas the United States spends about one-seventh.

The relation of U.S. research-and-development efforts to the rest of the world can also be seen in patents and published research articles. For example, from 1966 to 1971, patents to U.S. inventors averaged about 51,000 per year; by 1984 that had declined by nearly one-quarter. Inventors elsewhere in the world had averaged about 16,500 patents annually from 1966 to 1971, but this figure almost doubled by 1984. The annual number of scientific and technical articles by U.S. authors increased at a far slower rate than that of the rest of the world.

Enough of dreary numbers. The message is clear: the United States is losing ground both absolutely and relatively. In a scientific and technical age you can draw your own conclusion about the consequences.

[1]The international comparisons are adapted from *Science Indicators*—1985. U.S. Government Printing Office, #038-000-00563.4.

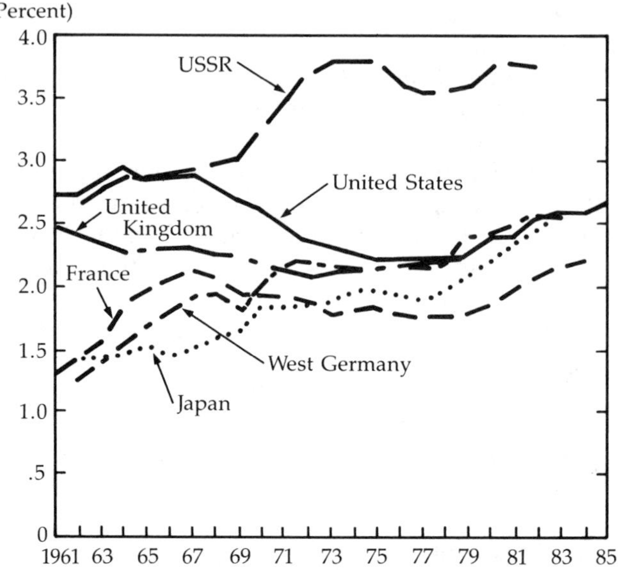

Figure 9-1. National expenditures for performance of research and development[2] as a percentage of gross national product by country. (From *Science Indicators*—1985)

Financing of Basic Research

In the last few years financial support for science has changed significantly. It is frustrating to report that, in spite of numerous statements about the extent and nature of these changes, it has been impossible to find sufficient agreement to be sure exactly what has happened. Two points do appear clear. First, there has been an overall cut in research funds, further reinforced by substantial inflation of prices. Second, there has been increased emphasis on "relevant" research—that is, research with immediate practical applications.

The following description of the support of research and associated activities in the United States relies primarily on the *Reviews of Data on Science Resources,* issued by the National Science Foundation (NSF). The figures don't represent any single year, since it was necessary to quote figures from different years (mostly 1982 and 1983) for various reasons. But they do present a general picture that, although dated, will give you some understanding of the situation.

[2]Gross expenditures for performance of research and development including associated capital expenditures except for the United States where total capital expenditures are not available. Estimates for the period 1972–1980 show that the inclusion of capital expenditures for the United States would have an impact of less than one-tenth of one percent for each year.

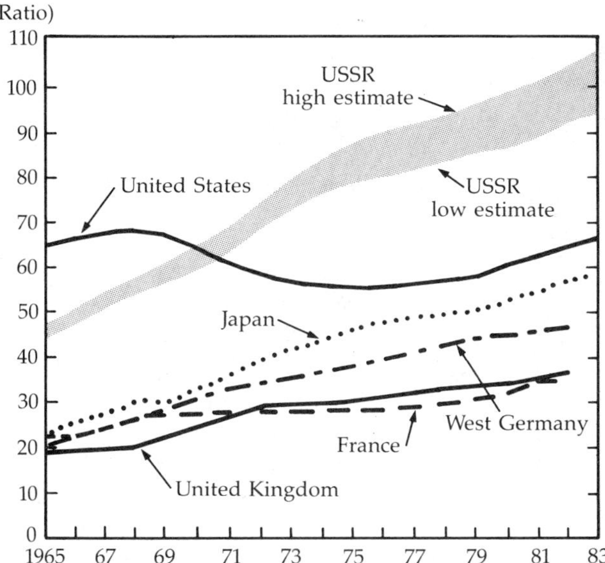

Figure 9-2. Scientists and engineers[3] engaged in research and development per 10,000 labor-force population by country. (From *Science Indicators*—1985)

Most reports published in popular media lump research and development funds together. This practice seriously distorts the picture and misleads the public about the research effort. The NSF uses three categories: basic research, applied research, and development. Consideration of a few representative figures will show the differences in these two approaches. When total research and development expenditures were approximately $106.6 billion, about $36.2 billion went into basic and applied research combined. Applied research accounted for about $22.9 billion, while basic research received about $13.3 billion. An analysis of several years' budgets shows that development got nearly two-thirds of the funds, whereas basic research received only about one-eighth. Development is primarily concerned with the technological applications of the sciences. In this sense it is a gross distortion to combine the research and development budgets, as is usually done in the popular media.

The distinctions between basic and applied science were pointed out in Chapter 2. Our primary interest as scientists is that part of the total expenditures concerned with basic research. These funds are the lifeblood of science in the modern world. Incidentally, both authors are convinced that the one-eighth figure for basic research is inflated and that approxi-

[3]Includes all scientists and engineers on a full-time equivalent basis (except for Japan, whose data include persons primarily employed). NOTE: A range has been provided for the USSR because of the difficulties inherent in identifying the relevant categories of Soviet personnel.

mately one-fifteenth is more accurate. Nevertheless, the amount (in constant dollars) devoted to basic research is quite large when compared with much earlier times. And it is only fair to state that the level at which basic research should be supported is a matter of judgment; also, the authors' personal interest in basic research is clear.

There are at least three points to be considered regarding the adequacy of the present U.S. effort in basic research. First, the absolute amounts are deceptive. The needs for funds have grown sharply, in part because of higher costs for new, more sophisticated and complex equipment. In the face of these needs, financing of equipment declined by about 41% in constant dollars from 1965 to 1979. Although we cannot locate any later figures, the evidence suggests that this condition continues to deteriorate. Considering the increased cost of equipment, the funds available will purchase less than half of what could be purchased in 1965. (We used the most conservative data; the real situation is likely to be much worse.) It may still be possible to advance basic science by flying a kite in a thunderstorm or by peering through a homemade microscope, but it is much more likely that advances will be made by the use of powerful particle accelerators and electron microscopes, operating with computer-controlled precision. Even at today's prices, Franklin's kite, made of a "large thin silk handkerchief," would cost only a few dollars, whereas construction of a powerful particle accelerator might cost several hundred million dollars. And yearly maintenance would require more millions. Franklin's letter to Peter Collinson describing his kite experiments took slightly over 400 words. The collection, storage, retrieval, and analysis of information collected in a single accelerator experiment require extensive use of specialists and additional complex computer facilities. In brief, science now operates on an entirely different scale from that of earlier times. The growth of costs in basic research is not in itself an argument for whether or not more funds are needed, but it is important information that has to be considered.

The second point regarding funds for basic research has to do with the adequacy of financing certain types of research. For example, in fiscal 1985 the social sciences received about $143 million from the federal government. This was to be spread over the disciplines of cultural anthropology, economics, history, political science, sociology, and any other research undertaken primarily for the purpose of understanding group behavior. Even though most social-science projects are not as expensive as physical-science projects, the amount allotted is only a stale crust. Social scientists investigate problems that are crucial to individual, national, and international well-being. Further prospects for adequate financing of widespread social-science research are not very cheerful.

Probably the major reason why the social sciences have been given so little attention is that their studies almost inevitably ruffle some representative's or special-interest group's sense of what is right and proper. Very few would understand or care about the implications if a new subatomic particle were found or the law of parity overthrown. An honest economist, sociologist, or political scientist can scarcely find a topic that does not outrage at least part of the populace. Those who are most likely to be outraged are defenders of the status quo, since serious research is unlikely to find that our social arrangements make this the best of all possible worlds.

On the other hand, it is sad to report that certain researchers in the social sciences have found ways to justify their role and do research that supports the controllers of the purse strings in Washington. This support tends to assure these researchers of continued money for their endeavors. How this works is spelled out in an interesting article by Alvin Gouldner (see the Bibliography).

The third point bearing on the adequacy of funds for basic research involves the support and training of future scientists. A substantial portion of the funds in any research program supports either graduate students as research assistants or junior scientists who work on particular projects. There is no substitute for apprenticeship training. The number of scientists trained for basic research depends on decisions made several years in advance, and the amount of funds allocated to basic research is one of the most important factors in such decisions.

The NSF's publication *Science Indicators* (1985) is our major source of information on combined private and public funds spent in the United States on basic research. About 83% of this total is from the federal government. These funds depend on the political climate. They increased relatively consistently from 1960 to 1968 but decreased from 1970 to 1976. From 1977 to 1985 there was an increase in support from both the federal government and industry. In terms of constant dollars the 1985 estimated total expenditures from all sources for basic research was $13.3 billion. This is 12.4% of the total research-and-development budget. The amount spent on basic research has been going up on the average of less than 3% a year for the last 15 years. Considering population and industrial growth, increased administrative costs, and the use of more sophisticated equipment, research support has been decreasing at a time when the United States is losing its scientific and technological advantage in the world.

It is interesting to note in passing that colleges and universities allocate well over half of their own research-and-development funds to basic research, whereas, according to a 1985 survey, industry allocates only about one-twentieth of its research-and-development funds to the same area. And this latter figure is probably overstated because of the glamour implicit in basic research.

The trends in funding appear to be due to several factors. First, Presidents Lyndon Johnson and Richard Nixon seemed to have little understanding of science. Gerald Ford and Jimmy Carter, however, were much more sympathetic to it. Ronald Reagan gave the greatest priority to military research. His appointments of science advisors and his view of nonmilitary science indicated neither sympathy nor understanding. The general attitude of Johnson and Nixon was that science is supposed to concentrate on immediate, usable results. Second, Johnson and Nixon gave evidence of a positive dislike for scientists. Johnson's attitude appears to have been related to widespread and open opposition to the war in Vietnam; Nixon was unpopular among scientists for a wide variety of reasons. Third, budgetary pressures have had serious effects on many programs. Fourth, although scientists are given a good measure of respect by the general public, they are also blamed for such things as threats of atomic war, a deteriorating environment, and the energy crisis. Fifth, many members of Congress view scientists with less than full favor; others feel that members of Congress are more capable judges of what is valuable and appropriate research than are scientists. This latter attitude could be clearly seen in an amusing (only in retrospect) and asinine bill that gave Congress the right to review individual NSF research grants. It soon became clear, even to Congress, that it did not have the technical capability to judge complex scientific studies. Whom could they consult? Scientists, of course. The death of this bit of nonsense should have been accompanied by some red faces. However, Senators Edward Kennedy and William Proxmire continue their often poorly informed badgering of science and scientists. One wonders whether either is aware of the cases of Galileo and Lysenko. Our national comedy is not always funny.

An article in *Science* by Philip Boffey told of a survey by the policy-making body for the NSF of a large number of research administrators at universities, industrial concerns, federal laboratories, and independent research institutes. These included university and corporate presidents, department chairmen, and laboratory directors. They showed strong agreement on a number of points, including the relative decline of science in the United States. Some quotations make the point clearly. From N. B. Havey, vice-president for research and patents of Bell Laboratories: "I would say that the single most critical issue with respect to long-term research in industry is that it is not being done, for the most part. A few companies in a few industries support it, but the bulk of industry has either given up or never did it." From Hans Mark, director of the Ames Research Center of the National Aeronautics and Space Administration (NASA): "I have noticed in the past eight or ten years a distinct drift of our very best people away from the basic fields. . . . The quality is not as good as it once was." From Harold Agnew, director of the Los Alamos Scientific Laboratory: "The ever increasing bureaucracy composed of managers who require more and more detail, justification, and guaranteed schedules, will in the not too distant future completely eradicate our nation's world position in research and technology."

Even assuming that these statements are somewhat gloomy, the picture is certainly not good. If, as we believe, basic research is at the core of modern thought as well as industry, continuation of present trends will further erode the U.S. position in world affairs.

Distribution of Federal Funds

Although there are serious problems concerning the level of financing, there are still funds available for basic research. How are these funds awarded to specific investigators? We will consider only the distribution of federal funds, since the federal government is the largest single source. The distribution practices of institutional, private, and industrial sources vary too widely to be discussed meaningfully.

The typical procedure in federal agencies is for the prospective investigator to submit a research proposal in some reasonably standardized form. A board of scientists evaluates the proposal and compares it with other proposals. The board of scientists is typically composed of men and women of some standing in their own discipline who serve as government consultants. In general the system of peer review has worked well. However, these boards tend to be quite conservative. For example, funding of established scientists has been virtually assured, while newly hooded Ph.D.s often have a rather difficult time unless they are personally known to some of the board members. Young Ph.D.s may have a particularly difficult time if they propose an innovation or departure from established thinking. Two possible solutions might be to earmark additional money for young investigators or to give funds to various institutions for local distribution.

There have been some recent, disturbing developments in allocation of funds. Congress has appropriated funds specifically for certain universities, and some of the college governing boards have become politicized. For example, there is the recent statement by the assistant secretary of defense for research that those who opposed "Star Wars" openly should not get research support from the Department of Defense.

There is a particular danger inherent in any system of research in which budgets are determined or funds allocated by administrators and/or committees of established scientists. This danger is that the direction of research will be where the money is. To some extent this does occur, and such a result degrades both science and scientists. In addition, since administrators and established scientists tend to work out existing problems rather than create new problems or venture into new fields, this system is capable of producing sterility.

Inevitably, there have been criticisms of federal research programs. Some of these criticisms are unquestionably justified. All programs need to be watched closely and criticized judiciously. Scientific programs are not exceptions. However, the critics ought to do their homework before launching their barbs. One of the best examples of tainted criticism was the lead editorial in a newspaper attacking a study of "mother love" in monkeys as an obvious waste of money. The wisdom of the editorial can be evaluated in the light of the fact that the scientist criticized, Dr. Harry Harlow, received a National Medal of Science award for this work—the federal government's highest award for distinguished achievement in science, mathematics, and engineering. Unfortunately, science has yet to discover a way to hybridize the "know-nothings" with the "say-nothings."

Three final points. First, according to the most recent data, approximately 83% of all federal research-and-development funds were controlled by three military-related groups: the Department of Defense, with 75%, the National Aeronautics and Space Administration, with 5%, and the Department of Energy (formerly the Atomic Energy Commission) with 3.5%. Obviously not all of these funds went to strictly military projects. Yet the simple fact remains that we spend vastly more learning to kill people than we do preserving and enriching life.

Second, unfortunately, and contrary to the belief of most people (apparently including Senator Proxmire), professors do not get rich on research grants. Generally, faculty members can draw only their regular salary or a portion of it through a grant. Research funds may relieve faculty members of some other duty, support them through the summer, or lead to future advancements, but the funds don't directly increase their monthly paycheck.

Third, since the federal government is the principal source of research funds, its policies regarding research are critical for the entire population, not just the scientific community. If the government decides to mount a war on cancer, as it has, the remainder of the research community pays for it. Further, just throwing money at a problem does not necessarily result in a solution, although we have had some major successes. Compare the results of the Manhattan Project (the development of the atom bomb), the Apollo moon program, and the cancer program. Each was directed toward solution of a problem. The Manhattan Project and the Apollo program were clearly successful; the cancer program's achievements have been much more modest. Why? One very important reason is that the basic knowledge for the Manhattan and Apollo programs was in hand or in sight. This cannot be said of the cancer program. As we commented in Chapter 2, it is difficult to put a time target on the acquisition of basic knowledge.

"Targeted" research often has the effect of forcing scientists to choose whether to work in an area in which they have only limited interest or forgo funded research. The problem affects more than just their research. It also influences education and training. Major fluctuations in these fields lead to instability and a great deal of waste in human and other resources.

The political attitudes that lead to this sad state of affairs are epitomized by Senator Kennedy's remark that science is too important to be left to the scientists. Science, obviously, should be done by politicians. The problem is that the relationship between science and politics cannot get too close without resulting in the degrading of science. We have presented this issue before: working directly on a practical problem is not usually the way to solve it if the solution requires new basic-scientific information. President Carter was more enlightened than his recent predecessors about scientific matters and temporarily improved the situation. President Ronald Reagan rarely supported any other than military research. Reagan's position that the creation story should be taught along with evolution lent little support to the idea that he understood or had any commitment to science.

Science and the Humanities

In recent years there has been considerable discussion of the relation between science and the humanities. C. P. Snow, the English scientist, novelist, and essayist, has presented the thesis that the intellectual community is divided into two "cultures," science and the humanities. Like other sweeping generalizations, this view seems to be both true and untrue. It would be a trifle devious to argue at this point that there are no differences in attitudes between the two. But one would have to be a bit thick not to see that there are also a number of similarities. The idea of two cultures is an oversimplification, if for no other reason than that there are many cultures, each of which may be subdivided into a number of variations. Still, it is worthwhile to ignore some of the finer points and consider the relation of science and the humanities.

The differences in attitudes are related in part to the different objectives. In gross terms, one objective of science is to achieve precise and parsimonious statements about the structure and processes of the animate and inanimate worlds. Ideally, these statements allow us to describe, understand, and predict something about those worlds. As stated earlier, elegance and aesthetic appeal have their place in the world of the scientist, but these qualities can be expressed in terms of precision and parsimony. Einstein expressed the objective briefly: "Supreme purity, clarity, and certainty at the cost of completeness."

Some of the disciplines that fall into the general category of humanities are poetry, art, drama, music, literature, and speculative philosophy. Other disciplines, such as architecture (project housing excluded), history, linguistics, ethics, and cultural anthropology have a foot in this camp. In any event, a primary objective of the humanities is to enrich the life of the beholder by arousing some sensual experience, emotion, or feeling. Some of these feelings are quite complex and intricate and need developing—an activity that requires a great deal of talent. Objectives such as precision and clarity are important in some cases; in others the effects may be achieved by complexity or a veiled hint. Granted that there are differences in objectives between science and the humanities, are there any necessary conflicts? The answer is an emphatic *no.* Ideally, the sciences and humanities should complement one another. Recall our plea for a broad education for scientists. This same plea can be extended to other disciplines as well. The devotee of the humanities who has no understanding of the general ideas of science is just as limited as the scientist for whom art forms arouse no emotion. Scientists don't have to be experts (although it may help) in order to have their feelings aroused by Hamlet's soliloquy, the prison scene from Gounod's *Faust,* a line drawing by Picasso, or Richard Wright's *Black Boy.* By the same token, the general thesis of this book is that science is not a mystery that can be confided only to the elect.

Some of the differences between the humanities and sciences undoubtedly involve the personalities of individuals who are attracted to the different disciplines. There is some evidence that particular areas tend to attract and hold people with certain personality characteristics. If this evidence is correct, these differences among people are likely to be emphasized and enlarged by their discipline.

There are a number of ways in which the humanities and sciences are alike. One of the principal likenesses is the motivations of the professionals. There are probably two primary motivations for both groups: (1) the participants have found their game, and (2) playing the game is important to them. In addition, both groups find in their game an escape from the crass, drab, everyday world. It is interesting to note that their products make the world somewhat less crass and drab.

There is a very practical problem that has led to an emphasis of the division between the humanities and sciences—the matter of financial support. Beginning with World War II there was substantial support for science, in great part due to the rather ignoble reason that science is necessary to war-making and international one-upmanship. The humanities have not enjoyed this support. There are clear indications of envy and ill feeling typical of the attitude of have-nots toward haves. To be sure, the federal government has recently taken some timid steps in the right direction, but funding of the humanities is so minimal that the program is hardly more than a sick joke. For example, the National Foundation on the Arts and Humanities spent only about $315 million for fiscal 1980—less than the cost of one bomber. It is some sort of commentary on our culture that we enthusiastically support the appropriation of billions for a war-making potential in order to preserve our "way of life" while appropriating only a pittance to make that "way of life" richer, more meaningful, and in general more deserving of defense.

Other nations have made equally valid decisions. For example, China, as a part of its "great leap forward," sent scientists, musicians, academics, and other intellectuals to farms and factories to learn humility. The results were an unmitigated disaster. They returned after a decade with at least some measure of freedom. Observers of China think it will take a generation to regain lost ground.

An argument has been raised that, if artists, writers, and others in the humanities begin to receive governmental support, they are likely to be controlled and directed. Science has remained largely independent of governmental control even though primarily supported by governmental funds. Good scholars and artists tend to be very independent. The danger of government domination exists in all research and work, including the social sciences. But even under authoritarian governments, such as the Soviet Union, Argentina, and South Africa, intellectuals have proved very difficult to control (Sakharov, for example).

Whatever the division or divisions between (and among) the sciences and humanities, there are two points worth considering. First, the amount of material in most disciplines is simply beyond the comprehension of any individual. In one particular field, approximately 29,000 books and articles were published in a year's time. Some other fields go far beyond this total. About 40,000 journals go to press regularly, and these journals contain over 24 million pages annually. The point is clear: it is extremely difficult to be well informed even in a single field. Universal scholarship like that possible during Aristotle's day does not occur. We cannot expect either scientists or their counterparts in the humanities to be well versed in many diverse fields, but the immensity and complexity of material is still no excuse for a lack of general knowledge. Second, there is great virtue in diverse approaches, even though diversity produces some confusion. The last great intellectual synthesis in Western Europe occurred in the Middle Ages and was enforced through the suppression of opposition. Intellectual stagnation was the price of uniformity. It seems obvious that pluralistic attitudes and objectives have the same virtues as a pluralistic society—variety, excitement, confusion, and the ability to change.

We worry about what the trends in our colleges and universities imply about the future of the United States. A few years ago Robert Nichols, program director of the National Merit Scholarship Corporation, surveyed the career decisions of semifinalists for the scholarship. "The overall trend is that the interest of able students in physical sciences and engineering has been decreasing during the period covered by this study, and that interest in the social sciences and humanities has been correspondingly increasing," he said. We know of no recent studies of National Merit semifinalists, but our informal surveys of student interest no longer bear this trend out. Many fewer students are showing interest in the physical sciences, humanities, or social sciences, while there has been an increase in interest in business and engineering. In the sciences, the drop in freshman majors was about 40% (about 21% if computer science is included) between 1974 and 1983. The big winners among freshmen with high grade point averages were engineering (+67%), business (+89%), and computer science (+580% from a very small base). Our impression is that students are being strongly influenced by the country's economic travails and are headed for financially secure careers. Historically, these disciplines have not been a major source of intellectual innovation.

10
The Scope of Science

The object of this chapter is to organize some of the matters that were discussed earlier. We view science as a whole and discuss both its breadth and its limits.

General Ideas and Attitudes That Have Shaped and Guided Science

The following is a broad description of how we think most scientists view science. If you really understand these ideas, you will begin to have a reasonable view of science. Some of the material in this section appears elsewhere in the book. We think the points are important enough for repetition. It has taken us more than one exposure.

1. Events are to be examined in terms of their natural physical causes. We could have spent several chapters explaining this, particularly the meaning of "causes." This general statement carries with it the implication that mystical or supernatural interventions are not useful as explanations. Many of the scientists and philosophers who helped develop this most important of ideas were professedly religious (Bacon, Newton, Copernicus, Kepler, Galileo, Kant, Aristotle, etc.). Others (Darwin, Franklin, Helmholtz, Russell, Hume, Crick, Spinoza, Huxley, Einstein, etc.) either were not religious, or

viewed religion and nature as one. Whatever their personal religious beliefs, their ideas led to the description and understanding of a universe without supernatural intervention. Today, there are almost no scientists who attempt to mix the supernatural with science.

2. Science rests on a base of observation. Certainly scientists often make conjectures or build models and theories before the data are available. They may even do this when the existing data seem to contradict their notions (consider Copernicus, Einstein, and Wegener), but science requires that important contradictions eventually be resolved.

Galileo played a crucial role in establishing this general principle, which is at the heart of the scientific enterprise. Even so, claims of the superiority of dogma, of whatever nature, have not been abolished. The dogmas of Lysenkoism (with Stalin presiding), creationism (Falwell, Gish, Morris, and others), and Aryan superiority (Hitler, the Ku Klux Klan, and others) have been proposed as scientifically valid, with no empirical justification whatsoever. Scientific concepts are not perfectly attuned to the data. We always have some things that don't fit and others that are ambiguous. Nevertheless, in the end we have to support our ideas with observations.

3. Processes in the universe are uniform. This idea is particularly dear to the hearts of physical scientists. The implication is that a gravitational field, for example, is always the product of masses and distances, whether we are talking about a football or a supernova. Some biologists (see Mayr's book, *The Development of Biological Thought*) have trouble with this general idea because biological organisms have particular genetic and experiential histories that affect their nature. They do not deny the universality of physical laws as regards inanimate objects or events, nor do they deny that these apply to certain aspects of the biological world. For example, the effects of gravity or radiation can be lawfully derived for camels or crickets. It seems to us that the point biologists make is this: complex organisms have to be independently studied because they contain an organization of physical properties that does not generally occur. The uniformity doctrine may be thought of as saying that, even among different complex organisms, events can be lawfully related whenever and wherever they occur.

4. The statements made in science are probabilistic. This assertion has at least three variants. First, in physics today it is generally accepted that many observed events are dependent upon seemingly random processes. Consider the particles given off by uranium. Out of a given mass of uranium we do not know which atoms will give off the particles or in which direction they will move. However, we can calculate with a high degree of accuracy the net effect.

A second variant of the general idea that scientific statements are probabilistic is that measures are inexact, so a given measure only has some probability of returning a given result. This is true for almost any measure. For example, if we measure a tabletop or the hem of a skirt, in feet, the result might be quite replicable. However, if we measure either in angstrom units, we would have a range of measurements. In most cases such variations are not large and usually don't create any practical problems. But they cannot be ignored by the scientist.

There is a more interesting measurement problem. Under certain conditions the simple act of measuring or observing may change what is measured or observed, so the measurement cannot be directly checked. The classic example is the position and velocity of an electron, which we discuss later in this chapter.

The third variant is that belief statements may be shown to be in error. They are tentative and therefore only probabilistically known. A good example was the long-held and well-tested belief that the "noble gases" (neon, krypton, xenon, and radon) would not form compounds. A young chemist, Neil Bartlett, discovered that a certain set of manipulations did indeed result in compounds.

There are other factors that lead us to conclude that scientific statements are probabilistic. Consider one final example. Our statements are based on past experience; logically we have no assurance that things will always remain the same. Rocks thrown by hand into the air come back down. This is a very consistent finding. Suppose the chances are one in five hundred trillion that the rock will continue to rise. You may never have observed that event, but you cannot "know" that it cannot happen. This is not, however, a recommendation that you stand and observe a "friend" throw rocks over your head.

5. A new and important way to understand the universe rationally is in evolutionary terms. This notion is not confined to biological evolution, which we are all aware of. For thousands of years it was generally accepted that the universe was unchanging. We should hardly be surprised that this was the case. After all, men could look daily at the mountains, rivers, and seas. There seemed to be cycles, such as the seasons or the phases of the moon, but these were regular, not unique. However, about 1500, Leonardo da Vinci recognized that seas had once covered parts of Italy; his conclusion was based on fossils found in the mountains. Geology began to become a science about 1788 with the publication of James Hutton's book, *New Theory of the Earth.* A very significant part of our modern view of the earth was formulated in the 1960s when Alfred Wegener's idea of continental drift was accepted (with the name changed to *plate tectonics*).

The idea of the evolution of the universe began to assert itself about 1913 when Vesto Slipher discovered that galaxies seem to be moving away from the earth at high speed. By 1929 Edwin Hubble confirmed the general idea of the expanding universe. The idea of a "Big Bang" was developed by Alpher, Hermann, and Gamow about 1948 and was strongly supported by Penzias and Wilson in 1965.

The point of this history, which omits much more than it includes, is that we have come to accept the idea that the form of the universe has changed drastically over an almost unimaginably long time (the processes such as gravity have not) and that it will continue to change. The evidence is overwhelming and comes from many different sources. No competing idea has any substantial support.

The Breadth of Science

Although we have concluded that scientific methods are broadly applicable, we have discussed the current breadth of science only briefly. The number of different paradigms and different kinds of events recently investigated by scientists is very large. The number of scientists and the knowledge they have built up and the information they have gathered are truly remarkable.

Technology has developed to such an extent that, with the assistance of millions of computers in the United States, we can process billions of times as much information as could be processed 20 years ago. This technology allows for many more kinds of operations and analyses than were feasible in the recent past. For these and other reasons, new sciences have been created with new bodies of information associated with them. Because of the vast expansion of knowledge, many scientists feel that they have to specialize narrowly to keep up in a given field. Specialization has two unfortunate limitations: individuals interested in more than one field are rarely able to explore these different interests, and scientists are often unaware of work from other fields that is relevant to their own work. In spite of these drawbacks, most scientists must do research on relatively narrow topics in order to make a meaningful contribution. As these topics accumulate their own bodies of data, theories, and methods, the scientist has to specialize even further.

New sciences often develop when a research area is seen to lie partly in each of two sciences. For example, see the discussion on sociobiology in Chapter 7. One current problem of many universities is that formal departmental structure may prevent students interested in one of these new sciences from getting the best education available. At present, such students must elect associated courses in all the relevant departments.

In the last few years the social sciences have come of age. Social scientists no longer have to consider themselves second-class citizens among other scientists. Because of the growing bodies of knowledge and new techniques for gathering information, sociologists, economists, political scientists, psychologists in some areas, and historians can take their place along with the natural scientists. The social scientists have difficulty doing certain experiments—many of their problems cannot be isolated in the laboratory, and ethical considerations also prevent some research—so they have developed some of their own techniques. In certain instances, applied and basic science merge into what has been called "action research." The value system of the scientist combines with the search for understanding; the research is done in a natural setting with the theoretical expectation that the results of the research will be goals that the scientist wishes to achieve in society. Program evaluation, as discussed in Chapter 7, describes some combined basic and applied approaches to science.

Considering the kinds of problems social scientists have, a compromise between theoretical and ethical considerations seems a valid resolution of the conflicts. However, isolation of variables in the laboratory, which allows unconfounded observation, is more likely to lead to greater understanding of the processes involved and to better description of relevant functional relations between the variables. Therefore, experimental investigations should be used whenever feasible.

Although limited in the experimental procedures available, certain social scientists in the last few years have been developing some formal theoretical approaches, such as game theory and computer simulation, which may lead to major breakthroughs if they have not already done so.

One social science that has shown remarkable productivity and popularity in the last few years is actually one of the oldest sciences—linguistics. Even ancient civilizations studied their languages and tried to formalize some of the facts about them. But with recent interest in generating sentences, with innovations in semantic and pragmatic analysis, and with new techniques of analyzing speech, new problems with new attempts at solution have emerged. Many scientific fields are appearing with linguistics at one focus; biological linguistics, psycholinguistics, sociolinguistics, and mathematical or computational linguistics are four of them.

Just about everyone is aware of the recent major accomplishments of the sciences associated with medicine. One of these, genetics, which includes recombinant DNA and genetic engineering as well as an analysis of corollary physiological functions, already generates many ethical problems, and these will increase when we learn to predict and control human heredity. But genetics is still a fascinating area of scientific study. Other areas are equally exciting. For example, the theoretical and practical problems associated with organ transplantation are of great importance in understanding human physiology.

Among the many different and interesting problems in the many sciences, each of which often seems more important than the others, is one science that is extremely new and relevant to just about every other modern discipline: computer science. Associated with this new science are some general problems relating to symbol manipulation, problem solving, artificial intelligence, the monitoring and modifying of systems (any systems), the rapid analysis of data, and the study of automata (machines). The computer itself is a scientific and technological discovery of tremendous importance. It has made many previously impossible activities routine. Whereas a person may be able to make a few thousand operations per day, a computer can make 100 million operations per second and make decisions at the same time. It receives, processes, and acts on information in billionths of a second, whereas the fastest simple actions of people take about one-tenth of a second. To understand and be able to use this tool wisely requires the special attention of some scientists. Computer science, however, is not simply the study of a machine. The development of the computer has pointed to natural problems of considerable interest, such as the differences among languages and the differences between natural and artificial languages, the theory of problem solving, the theory of automata, memory systems, information processing, and perceptual systems.

Scientists in all fields as well as a variety of individuals such as bankers and production line workers will find the computer an important adjunct. Scientists and nonscientists alike should have at least a nodding acquaintance with it.

Science, as it develops and advances, is being carried into a myriad of areas, and the number of areas and paradigms is multiplying rapidly. Science not only has extended into many new realms of the natural world but is extending into many aspects of the man-made world, such as computers, business organizations, social institutions, and plastics. Many conceivable topics are not discussed in science, but very few topics can automatically be excluded.

The Limits of Science

As we have seen throughout this book, many important aspects of the way we live are affected by science. The environment in which we live has been radically changed by science and technology. Our attitudes, beliefs, and even ethical principles are influenced by scientific knowledge. However, even though science has these widespread influences and affects all aspects of our life, it has important limitations. We have looked at a few problems for which science has no answers. Now we should discuss the limits of science as a topic in itself.

The most important limitations of science are an integral part of its methods. There are three ways these limitations can be clearly seen: first, both data and statements about those data are only probabilistic; second, there are both technical and theoretical limits to the accuracy of measurement; third, whatever the accuracy of our measures, we cannot measure all aspects of a given event.

As stated in an earlier chapter, information about the real world ultimately has to be known through induction, and nothing that is known by induction is known for certain. There is no way one can gain certain knowledge about the real world. In the end, scientific statements stand or fall by induction, and it is inductive methods that have led to, and provided tests for, atomic fission and fusion theory, theories of natural selection, and computers. Scientists may or may not believe that their statements will be shown to be ultimately true, but if one accepts the concept of induction, scientific statements can be demonstrated to be reasonable.

Another limit of science concerns the accuracy of measurement. This accuracy is limited in a number of ways. There are limits on measurement due to technology. Intelligence and socioeconomic status are important variables, but we can't measure them accurately; we can't measure the weight of a gas to a great degree of accuracy; we can measure the speed of light only within limits; and so on. We can't measure anything completely accurately, and we never will be able to, if our current conceptions of physics are accurate. There is a principle in physics, known as the Heisenberg uncertainty principle, which leads to the conclusion that there are limits to the accuracy of our measures. When subatomic particles are involved, there is no way to measure simultaneously both their velocity and their location. Thus, one can't accurately predict a particle's future location and velocity. Furthermore, the apparatus used to measure subatomic particles affects their location and velocity; thus, the measuring instrument affects the object measured. In certain psychological and sociological measurements, there are different but parallel effects. Sometimes the test people take or the interview they undergo affects their behavior. One may not then be able to identify what the performance would have been had they not taken the test. One other limit on the accuracy of measurement should be mentioned. Certain measurements take time to collect, particularly in the social sciences—for example, gross national product or data on absenteeism. If the data are collected in a dynamic system, then they do not necessarily represent the current situation by the time they are collected.

In science, as well as in philosophy and other disciplines, unanimous agreement is rare. This is particularly true when an element of judgment is involved. Nearly all scientists would agree, for example, that there are limitations in measurement. The following paragraphs on the relation of science to values would probably be acceptable to most scientists and to a great many, if not most, philosophers. In any case, these are our views.

We can never know everything in science. Any specific event in the real world can be analyzed or categorized in an unlimited number of ways. Each of these has a very large number of variables affecting it. Since there are essentially an unlimited number of events, and only a small number of scientists with a limited amount of time and facilities, there will always be a great deal left to be investigated. The known world is incomplete and will remain so. We will never understand more than a small part of it. And since the ability of any scientist to conceptualize is limited, and the number of aspects of events unlimited, conceptualizations have to be incomplete.

We are limited in what we know of some events because, although we are attempting direct measurement, there is a limit to the accuracy of our measurement. Our knowledge of other events is further limited because there is no way to make direct measurements. In these cases, we have to infer the state of these events from indirect measurements. Consider human values; they are what make life worthwhile (that life is worthwhile is itself a value judgment). On what basis do we decide that something is good? Beautiful? Fun? Interesting? Psychologists may be able to answer each of these questions. They may be able to find out what we *think* is good, or beautiful, or fun, or interesting, and they may eventually be able to tell us why we think so. They will come to these conclusions by indirect measurement. But in *no* instance will they be able to tell us whether something is *really* good, beautiful, fun, or interesting. You may object, saying that something may *be* really good or really beautiful, that something is *automatically* fun or interesting if we enjoy it. You may argue that fun, by its very nature, cannot be analyzed any further. Your argument may or may not be valid. It is irrelevant. The point is that a scientist can do the same kind of investigation with any of these values. Going beyond these limits to say whether something is *really* good or *really* beautiful is not within the realm of science.

These topics undoubtedly need to be analyzed, and philosophers do, and should, investigate the question of absolute good, just as psychologists and other social scientists investigate what people think and believe about goodness. But there are limits to any of these analyses. A philosophical analysis may show what logically follows from certain assumptions about values, and these results may then be evaluated, but nowhere will the analysis state "something is good" with certainty. A theological, religious, or mystical source may indeed state that "something is good" and one may feel certain of it, but there is no scientific evidence to justify the feeling of certainty. All that anyone can do is to decide on some intuitive basis that something is good and (perhaps) act accordingly. To help their intuitions and to help them make these decisions about goodness, scientists may use the knowledge of the consequences of certain activities. That is, they may evaluate the consequences as well as the activities.

Exactly the same arguments hold for other values; scientists can only find out what people believe and the consequences of such beliefs; they cannot demonstrate the truth of such beliefs.

The last "limit of science" we will discuss falls under the general heading of "metaphysics." These questions are about reality directly. Not: Is it really good? Is it really beautiful? But: What *is* it really? What are electrons? What is a table really made of? Do minds really exist? Do people have immaterial souls? Does magnetism exist? These questions can be most exasperating. And they cannot be answered. Scientists start with a set of undefined terms. They may think that they are real, and they may give them some properties, but they can't tell you what they really are. They may not even understand the question. Scientists attempt to confirm a theory in terms of observations. They can specify the conditions for observation, and they can describe the properties of the events observed.

They can relate the events to other events in a meaningful way. Those are the limits of their ability. Due at least in part to the limitations of science, some people turn to other areas in an attempt to get true and certain information. They may turn to authority, intuition, superstition, psychedelic drugs, or religion. These sources may give an individual the feeling of certainty, but there is no logical reason why a feeling of certainty is any more likely to be true than a belief without the accompanying feeling. Some data even suggest that feelings of certainty are often accompanied by more errors than feelings of uncertainty. Any system stating that something is true by fiat, authority, or vision cannot demonstrate its truth. People have no reason to accept it as true whether they are certain that it is or is not.

In conclusion, it can be seen that there is no part of the observable world that is automatically immune to scientific investigation. Even those personal aspects of human existence that "feel" immune to science can be studied scientifically. There are no natural events that negate such study. Limitations, where they exist, refer either to measurement limitations, to ultimate solutions, or to values. We all hope that the ethical principles followed by scientists are those leading to "life, liberty, and the pursuit of happiness," and we also all hope that scientists consider both their actions and the consequences of their actions in terms of human welfare. But, since ethical judgments ultimately rest on an ultimate value, they are basically immune to scientific investigation. Only the events as such can be understood scientifically.

A Few Concluding Remarks

Four principal themes have been followed in this book. Each one appears in several sections and perhaps deserves re-emphasis.

Science has been viewed as a game not because it is trivial or frivolous, but because it has many of the attractive elements associated with games. Monkeys will work long periods of time solving puzzles, and most scientists are fascinated by intellectual puzzles. Whether monkeys and scientists should be grouped in this way is a matter for speculation, but the evidence is clear that working on puzzles is intrinsically rewarding to both groups.

Beginning in Chapter 2, we emphasized the importance of attitudes. There are many possible ways of analyzing or examining any object or event. The attitudes and values of scientists set the boundaries of what is an acceptable statement or item of evidence within science. The emphasis is on objective and verifiable evidence and statements that are clear, logical, and at some point related to or confirmable by production of evidence.

Discussions of science have sometimes revolved around the scientific method. We take the view that, while there are methods within sciences, there is no single scientific method. This is not to deny the importance of methods and procedures, but rather to stress that methods and procedures are related to criteria for acceptable hypotheses and evidence. For example, social behavior of nonhuman primates is a topic of current interest for ethologists and psychologists. Ethologists typically favor field studies in which animals are studied in their natural habitat. Psychologists are more likely to study such behaviors in a laboratory setting. Each method has advantages and disadvantages. Whatever the method used, the common objective is to secure reliable information. Methods that do not lead to reliable information would be rejected by both ethologists and psychologists.

In several sections we have attempted to outline both the scope and limitations of science. Although there are limitations to science, we must not overlook the fact that science is extremely broad and far-reaching in scope. Complex but understandable. Fallible, but capable of self-correction. Restricted, but with fantastic capability for growth. Imperfect, but struggling. And, of course, it's fun.

Annotated Bibliography

The books and articles cited below were the most direct references used in the writing of this book. We did not usually refer to them directly in the text, since we did not usually present specific information having only a single source and since we wanted to make the book more readable by omitting footnotes. We annotated the bibliography to help you find books and articles about science that may interest you.

American Scientist. The general journal of Sigma Xi, the Scientific Research Society. Articles on a variety of scientific topics; usually can be understood by those with a limited scientific background. Strongly recommended.

Asquith, P. D., and H. E. Kyburg. 1979. *Current Research in Philosophy of Science.* East Lansing: Philosophy of Science Association. As usual with collections of articles some good, some . . . We particularly like the article by David Hull in which he considers the nature and role of philosophy of science in general and applications to biology more specifically.

Barnett, L. 1957. *The universe and Dr. Einstein.* New York: New American Library. A popular presentation of the view of the world according to modern physics; well written and clear. (Some very questionable opinions about Einstein)

Blackett, P. M. S. 1967. The ever widening gap. *Science, 155:* 959–964.

Bohr, N. 1961. *Atomic physics and human knowledge.* New York: Science Editions. A great scientist looks at atomic science and its implications for other aspects of science and humanity.

Boring, E. G. 1950. *A history of experimental psychology.* 2d ed. New York: Appleton-Century-Crofts. A fairly difficult and detailed but well-written history. A good discussion of the concepts and people that led to modern psychology.

Boronowski, J. n.d. *The common sense of science.* New York: Vintage Books. An excellent book on science by a mathematician who is very much aware of other aspects of our culture. Good chapter on "The Idea of Chance."

Britannica book of the year. Chicago: Encyclopaedia Britannica. This includes *World Data,* which contains statistics from countries around the world. Our infant mortality data are derived from this source.

Broad, W., and N. Wade, 1982. *Betrayers of the truth: Fraud and deceit in the halls of science.* New York: Simon & Schuster. We have mixed feelings about this. On the one hand, it is good to remind scientists and the public of human limitations and failings (even though they identify a historian as one of their scientists). On the other hand, their concentration on dishonesty may overestimate its incidence. Their "cures" don't seem to offer much new. We do recommend that you read the book—but cautiously.

Brown, H. I. 1977. *Perception, theory and commitment: The new philosophy of science.* Chicago: University of Chicago Press. A well-constructed presentation of some of the issues leading to the new philosophy of science.

Bunker, J., B. Bunker, and F. Mosteller, eds. 1977. *Costs, risks and benefits of surgery.* New York: Oxford University Press. An extensive consideration of how to evaluate the effectiveness, costs, and benefits of both new and established surgical procedures. The techniques and problems of evaluation receive extensive attention. Seems likely to be the standard in this field.

Campbell, D. M., and J. C. Higgins, eds. 1984. *Mathematics: People, problems, results.* Vols. 1, 2, and 3. Belmont: Wadsworth. For the most part these do not require a high degree of mathematical sophistication. They do give a good insight into the world of mathematicians.

Carnap, R. 1966. *Philosophical foundations of physics.* New York: Basic Books. A cogent presentation of philosophy of science by an eminent philosopher. The book was edited by the well-known writer Martin Gardner.

Christopher, M. 1970. *ESP, seers and psychics.* New York: Thomas Y. Crowell. A survey of many examples of psychic and mystical experience by a well-known magician.

Cipolla, C. M., and D. Birdsall. 1979. *The technology of man.* New York: Holt, Rinehart & Winston. Profusely and well illustrated, it gives a real sense of the long climb from chipped pebbles to modern technology.

Comroe, J. H., and R. D. Dripps. 1976. Scientific basis for the support of biomedical science. *Science* 192: 105–111. A report of some very important research into the relative impact of basic and applied research on cardiovascular and pulmonary problems. A model for investigating other "relevant" research.

Conant, J. B. 1951. *Science and common sense.* New Haven: Yale University Press. A very readable and clear attempt to give the nonscientist an understanding of science.

Conant, J. B. 1957. *Harvard case histories in experimental science.* 2 vols. Cambridge: Harvard University Press. Details of different specific episodes in the development of the natural sciences.

Cournand, A. 1977. The code of the scientist and its relationship to ethics. *Science* 198: 699–705. A fresh look at the pressures on scientists and a code of conduct.

Cracraft, J. 1983. *The Soviet Union today.* Chicago: University of Chicago Press. Originated in a series of articles published in *The Bulletin of the Atomic Scientists.* A rational look at the other superpower. It covers many aspects of Soviet life including science and technology. Strongly recommended.

Crease, R. P., and C. C. Mann. 1986. *The second creation.* New York: Macmillan. An account of the attempt by physicists to develop a unification theory that would give an overall model of physical processes in the universe. Attention is given to work in quantum physics. Most fascinating are the formal and informal pictures of the scientists involved, their errors, confusion, insights and triumphs. Well done.

Darwin, C. 1859. *Origin of Species.* New York: Modern Library. What can we say? The "Historical Sketch" alone is worth the price of the book.

Davies, J. T. 1973. *The scientific approach.* 2d ed. New York: Academic Press. An excellent survey of science as a general topic, by what is a rare bird in the United States: an educated engineer (educated in England, of course). Remarkably parallel to *The Game of Science* in many respects.

Davis, P. J., and R. Hersh. 1981. *The Mathematical Experience.* Boston: Houghton Mifflin. The kinship of problems of scientists and mathematicians is well expressed; ". . . mathematical knowledge as it really is—fallible, corrigible, tentative and evolving, as is every other kind of human knowledge." So much for certainty.

Dukas, H., and B. Hoffman. 1979. *Albert Einstein: The Human Side.* Princeton: Princeton University Press. Excerpts from letters, talks, articles, etc., by two who knew him well. Delightful.

Einstein, A. 1934. *Essays in science.* New York: Philosophical Library. Translation of a number of short essays. Most of these are readable with little or no scientific background.

Einstein, A. 1954. *Ideas and opinions.* New York: Crown. More short and mostly readable comments and essays.

Eldredge, N. 1982. *The monkey business.* New York: Washington Square Press. A well-written and readable examination of evolution and the attack by creationists. Includes an excellent short account of the differences between science and nonscience.

Eldredge, N. 1985. *Time frames: The rethinking of Darwinian evolution and the theory of punctuated equilibria.* New York: Simon & Schuster. A clearly written evaluation of the state of evolutionary theory together with a discussion of the Eldredge and Gould model, which may revolutionize the field.

Erlich, P. R., C. Sagan, D. Kennedy, and W. D. Roberts. 1984. *The cold and the dark: The world after nuclear war.* New York: Norton. Discusses the possibility and consequences of nuclear winter. A chilling must.

Fallows, J. 1980. American industry: What ails it, how to save it. *Atlantic* 35–54. An excellent and interesting look into the problems of productivity and industrial success. The answers are not clichés. Research is, at best, a part of the story.

Feigl, H., and M. Brodbeck, eds. 1953. *Readings in the philosophy of science.* New York: Appleton-Century-Crofts. A large book of readings covering many aspects of philosophy of science, ranging from relatively easy to quite technical and difficult.

Feigl, H., and W. Sellars, eds. 1949. *Readings in philosophical analysis.* New York: Appleton-Century-Crofts.

Feigl, H., W. Sellars, and K. Lehrer, eds. 1972. *New readings in philosophical analysis.* New York: Meredith. Conceptual analysis has become more important in scientific theorizing. These books have many classical papers.

Feyerabend, P. K. 1975. *Against method.* London: New Left Books. A well-written presentation of current radical philosophy of science by one of its most creative supporters. His analyses of the history of science are worth reading. We believe he overstates the role of the irrational in science and reaches some rather bizarre conclusions.

Feynman, R. P. 1985. *Surely you're joking, Mr. Feynman.* New York: Norton. A genuinely funny book by a Nobel Laureate with very little mention of his field. We suggest that it be made into a movie titled "A day at the physics lab" starring the Marx brothers.

Franks, F. 1982. *Polywater.* Cambridge: MIT Press. An excellent telling of the tale of a discovery that never existed, although about 200 researchers believed in it enough to work on it. The results are now considered to be the result of impurities. Another case of dirty test tubes.

Gardner, H. 1983. *Frames of mind: The theory of multiple intelligences.* New York: Basic Books. An interesting and far-ranging discussion of the limits of the standard methods of evaluating intelligence and some proposals for broadening the concept.

Gardner, H. 1985. *The mind's new science: A history of the cognitive revolution.* New York: Basic Books. A discussion of cognitive science and a review of its history.

Gardner, M. 1957. *Fads and fallacies in the name of science.* 2d ed. New York: Dover. A brief look at pseudoscience.

Gee (Gerhardt), J., and I. Savasir. 1986. On the use of Will and Gonna: Toward a description of activity types for child language. *Discourse Processes* 8: 143–175.

Gilbert, S. P., R. J. Light, and F. Mosteller. 1975. Assessing social innovations: An empirical base for policy. In Bennett, C. A., and A. A. Lumsdaine, eds. *Evaluation and experiment: some critical issues in assessing social programs.* New York: Academic Press. A review of evaluations of social, medical, and sociomedical innovations. Since most don't work, the necessity for good evaluation is emphasized.

Gingerich, O., ed. 1975. *The nature of scientific discovery.* Washington: Smithsonian Institute Press. The printed version of a symposium commemorating the 500th anniversary of Copernicus's birth. Papers and discussions of science past and present by a group of generally outstanding historians, philosophers, and scientists. In general, quite good. Some nonsense.

Goldberg, S. 1984. *Understanding relativity.* Boston: Birkhauser. The change from the Newtonian mechanistic to a relativistic framework together with a description of the work on relativity. Some very astute comments on science in general. We used it as an important source on Newton.

Goldsmith, D., ed. 1977. *Scientists confront Velikovsky.* Ithaca: Cornell University

Press. Scientists consider many of the claims of Velikovsky and evaluate them against the evidence.

Gould, S. J. 1977. *Ever since Darwin.* New York: Norton.

Gould, S. J. 1980. *The panda's thumb: More reflections in natural history.* New York: Norton.

Gould, S. J. 1981. *The mismeasurement of man.* New York: Norton.

Gould, S. J. 1983. *Hen's teeth and horse's toes.* New York: Norton. Gould is an excellent essayist and interpreter of the theory of evolution. Although some of his interpretations may be controversial, they are always reasonable.

Gouldner, A. W. 1968. The sociologist as partisan: Sociology and the welfare state. *American Sociologist* 103–115. A readable discussion of the interaction of values held by scientists and their science.

Gregory, M. S., A. Silvers, and D. Sutch, eds. 1978. *Sociobiology and human nature.* San Francisco: Jossey-Bass. A collection of articles praising, damning, and analyzing sociobiology. The article by S. W. Washburn and F. Beach and the one by D. Hull appeared here originally.

Grun, B. 1975. *The timetables of history.* New York: Simon & Schuster. Parallel listing of the events in history, politics, literature, arts, science, technology, etc. A very handy and useful reference, although there are some surprising omissions.

Hall, A. R. 1980. *Philosophers at war.* New York: Cambridge University Press. The dispute over priority in the invention of calculus by Newton, Leibniz, and their followers. Great men, scoundrels, provocateurs all play their parts.

Hansen, N. R. 1958. *Patterns of discovery.* New York: Cambridge University Press. An interesting and readable presentation of some aspects of the paradigmatic approach to science from a slightly different orientation. It is our major source on Kepler's research.

Harvey, W. [1628] 1962. *On the motion of the heart and blood in animals.* Chicago: Henry Regnery. An early presentation in the game of science.

Hempel, C. G. 1966. *Philosophy of natural science.* Englewood Cliffs: Prentice-Hall. An elementary presentation of philosophy of science by an eminent authority.

Hilderbrand, J. H. 1957. *Science in the making.* New York: Columbia University Press. A delightful little book by a practicing scientist. To be commended for insight (since we seem to be in agreement on most matters).

Hilgard, E. R. 1987. *Psychology in America: A historical survey.* New York: Harcourt Brace Jovanovich. A solid presentation by an eminent scholar.

Huff, D. 1954. *How to lie with statistics.* New York: Norton. A humorous little book that presents ways people can misrepresent data to give impressions that aren't valid. It tells what to look for in evaluating reports of data.

Hull, D. L. 1973. *Darwin and his critics.* Chicago: University of Chicago Press. Appropriately subtitled "The reception of Darwin's theory of evolution by the scientific community." Articles, reviews, letters, etc., by both proponents and critics of Darwin together with Hull's introduction and commentaries. As usual, well done by Hull who, unlike most philosophers of science, concentrates mostly on biology rather than physics or chemistry.

Hull, D. L. 1978. Altruism in science: A sociobiological model of cooperative behavior among scientists. *Animal Behavior* 26: 685–697. According to Hull, science works and scientists behave selflessly not because they are somehow superior, kind, and honest. Rather, their own self-interest coincides with the manifest goals of science. Whether you agree or not, the style and wit are worth the trip.

Hunt, M. 1982. *The universe within.* New York: Simon & Schuster. A reasonably articulate introduction to cognitive science written for the general reader.

Issues in science and technology. Published quarterly by the National Academy of Sciences, Washington, D.C. Covers a wide variety of topics and viewpoints. Generally gives overviews of important issues by outstanding scientists and others. The debates are usually spirited and informative.

Keller, E. F. 1983. *A feeling for the organism: The life and work of Barbara McClintock.* San Francisco: W. H. Freeman. The life, times, trials, tribulations, and triumph of a great biologist. One winces in realizing how shabbily women have been treated by male scientists.

Killian, J. R. 1977. *Sputnik, scientists and Eisenhower.* Cambridge: MIT Press. Killian was the first special assistant for science and technology to the president. An insider's view of the politics, ideas, struggles and people at a critical juncture in U.S. science.

Kistiakowsky, G. B. 1976. *A scientist at the White House.* Cambridge: Harvard University Press. An interesting set of excerpts from his private diary. As the successor to Killian as science advisor to Eisenhower, he participated in decisions on missiles, arms reduction, and test bans.

Kitcher, P. 1982. *Abusing science: The case against creationism.* Cambridge: MIT Press. Another of the books countering the creationist nonsense. More sophisticated about the nature of science than most such endeavors.

Kitcher, P. 1985. *Vaulting ambition.* Cambridge: MIT Press. A philosopher of science examines sociobiology. Probably best used in graduate courses. His discussions of theory have influenced some parts of the game.

Koestler, A. 1971. *The case of the midwife toad.* New York: Random House. An interesting story of a scientist's tragedy.

Kohler, W. 1947. *Gestalt psychology.* New York: Liveright. Max Wertheimer, the founder of Gestalt psychology, never wrote a survey of it. This lucid book was written by his colleague and Gestalt's cofounder.

Kovacs, E. 1964. *Biochemistry of poliomyelitis viruses.* New York: Macmillan. Used as the primary basis of our short history of virology. Very interesting, but many parts of the book require special knowledge.

Kuhn, T. 1962. *The structure of scientific revolutions.* Chicago: University of Chicago Press. The book that got us started. An essay in the history of science. Kuhn introduced the notion of the paradigmatic character of science here.

Lakatos, I., and A. Musgrave, eds. 1970. *Criticism and the growth of knowledge.* New York: Cambridge University Press. A discussion of Kuhn's ideas by several eminent philosophers.

Langmuir, I. 1953. *Pathological science.* Colloquium at the Knolls Research Laboratory, December 18, 1953. This amusing and tragic transcript was made by R. N. Hall from a disc recording furnished by the Manuscript Division of the Library of Congress. Langmuir reports a series of findings, primarily by physicists, in which they found things that were never there. Interestingly, some of these findings generated hundreds of supporting studies.

Latour, B., and S. Woolgar. 1979. *Laboratory life: The social construction of scientific facts.* New York: Sage. An inside view of life as it existed in the Salk Institute.

Lecky, W. E. H. [1865] 1955. *The rise and influence of rationalism in Europe.* New York: Brazillier. This is a fascinating, if somewhat repetitious, account of the development of modern attitudes.

MacDougall, C. D. 1983. *Superstition and the press.* Buffalo: Prometheus Books. A frightening look at the gullibility (or pandering to public tastes) by some of our leading newspapers.

MacIsaac, D. 1976. *Strategic bombing in World War Two.* New York: Garland. A very readable account of a very difficult evaluation project.

Mayr, E. 1972. The nature of the Darwinian revolution. *Science* 176: 981–989. A searching analysis of "perhaps the most fundamental intellectual revolution in the history of mankind." Why was it so long delayed, and what was its scientific and nonscientific impact?

Mayr, E. 1982. *The growth of biological thought.* Cambridge: Harvard University Press. For all biologists, a must from one of the grand old men of science. The first 146 pages are also important to anyone interested in science or philosophy of science.

Medvedev, Z. A. 1969. *The rise and fall of T. D. Lysenko.* Trans. I. M. Lerner. New York: Columbia University Press. The sad and inspiring story of Russian scientists' struggle against brute force and dictated science. At his trial, N. I. Banilov declared "We shall go to the stake, we shall burn, but we shall not renounce our convictions."

Medvedev, Z. A. 1978. *Soviet science.* New York: Norton. The good, the bad, and the ugly in Soviet science. Many of the scientists are very good. The bureaucrats and administrators—well, they're just bureaucrats and administrators.

Medwar, P. B. 1984. *The limits of science.* New York: Harper & Row. A very clear discussion of science by one of its most distinguished practitioners. It's not as limited as the title suggests.

Morick, H., ed. 1972. *Challenges to empiricism.* Belmont: Wadsworth. A series of articles on philosophical problems associated with empiricism. Quite good!

Morrison, Philip, and Phylis Morrison. 1982. *Powers of ten: A book about the relative sizes of things in the universe and the effect of adding another zero.* San Francisco: W. H. Freeman. This book helps convey to the student the range of things scientists investigate. The plates are particularly impressive.

Mosedale, F. E., ed. 1979. *Philosophy and science.* Englewood Cliffs: Prentice-Hall. A rather interesting set of papers on the relationship between science and philosophy in areas such as religion, human nature, limitations and methodology, ethics, and society.

Mueller, R. A. 1980. Innovation and scientific funding. *Science* 209: 880–883. How one scientist succeeded and was awarded prizes for innovative work in spite of the bureaucracy. However, the bureaucracy is gaining and with a little luck should be able to stifle all innovation.

Needham, J. 1954. *Science and civilization in China.* New York: Cambridge University Press. In several volumes (11 at last count), the wide range of Chinese science, technology, and medicine is considered in depth. There is a later condensed version. (*The grand titration.* 1969. Toronto: University of Toronto Press.) Needham believes Chinese innovation and influence on the West has been greatly underrated.

Newman, J. R., ed. 1956. *The world of mathematics.* 4 vols. New York: Simon & Schuster. An interesting series of papers on the nature of both applied and pure mathematics. Many classic papers are included.

Norman, D. A., ed. 1981. *Perspectives of cognitive science.* Norwood: Ablex. Papers presented at the 1st annual meeting of the Cognitive Science Society, La Jolla, CA, August 1979.

Pais, A. 1982. *The science and the life of Albert Einstein.* New York: Oxford University Press. This has become a classic. Some of the technical material is tough without a background.

Piel, E. J., and J. G. Truxal. 1975. *Technology: Handle with care.* New York: McGraw-Hill. Faces some of the problems and dilemmas in applying science in the modern world. Well worth your time whatever your field.

Popovsky, M. 1979. *Manipulated science.* Garden City: Doubleday. One view of the strangling influence of Soviet bureaucracy on scientific efforts. It's even worse than in the United States.

Popper, K. R. [1934] 1961. *The logic of scientific discovery.* Reprint. New York: Science Editions.

Popper, K. R. 1963. *Conjectures and refutations.* New York: Harper & Row. A very important and influential book in the philosophy of science. Some of it is very tough going but worth the effort.

Randi, J. 1975. *The magic of Uri Geller.* New York: Ballantine. One famous magician investigates the "psychic powers" of another. He answers in part how it's done.

Randi, J. 1982. *Flim-flam! Psychics, ESP, unicorns, and other delusions.* Buffalo, NY: Prometheus Books. A detailed investigation of several famous cases.

Ravetz, J. 1971. *Scientific knowledge and its social problems.* New York: Oxford University Press. Contains a section on critical science reprinted in *New Scientist* and *Intellectual Digest.* Proposes that substantial scientific effort be directed toward examining the effects of scientific and technological developments.

Reid, T. R. 1984. *The chip.* New York: Simon & Schuster. A history of one of the most important scientific and technical developments of the 20th century—the microchip. Jack Kilby and Robert Noyce's results are known to and have an effect on practically everyone. Yet they are not recognized by a public to whom Edison and Bell are household names.

Rhine, J. B. 1953. *New world of the mind.* New York: Apollo.

Rhine, J. B. 1967. *ESP in life and lab: Tracing hidden channels.* New York: Macmillan. Whether you happen to be interested in this sort of thing or are just idly curious, these books contain interesting discussions of ESP from the viewpoint of ESP researchers.

Richter, J. P., ed. 1970. *The notebook of Leonardo Da Vinci.* New York: Dover Publications 2 vols. Illustrations and text from the famous notebooks. The text has the commentaries in both Italian and English. Of interest to both scientists and artists.

Roe, A. 1953. *The making of a scientist.* New York: Dodd, Mead. The author has done more than any other single individual to study the scientist as a person. Other monographs and articles by this author are recommended.

Rossi, P. H., and W. Williams, eds. 1972. *Evaluating social programs.* New York: Seminar Press. A series of papers on the uses of and problems in evaluation research.

Rossiter, M. W. 1982. *Women scientists in America: Struggles and strategies to 1940.* Baltimore: The Johns Hopkins University Press. A review of women's efforts to break into the scientific world. Enlightening to all, embarrassing to all men.

Russell, B. 1953. *The impact of science on society.* New York: Simon & Schuster. The title introduces the wide scope. In spite of some flaws in the assumptions and predictions, well worth reading and considering.

Sachs, M. A. 1971. Resolution of the clock paradox. *Physics Today* 11–17. An interesting discussion of a current controversy in physics.

Sayre, A. 1975. *Rosalind Franklin and DNA.* New York: Norton. James Watson acknowledges that Rosalind Franklin contributed greatly to the discovery of the DNA molecule. Anne Sayre gives a more accurate description of the woman and her contributions.

Schwartz, G., and P. W. Bishop. 1958. *Moments of discovery.* New York: Basic Books. 2 vols. Short sketches of the life and activities of a wide variety of scientists, beginning with Hippocrates. Short selections of their translated works are of particular interest.

Science. Published by the American Association for the Advancement of Science. Presents articles from many fields and on a wide range of levels. Strongly recommended.

Scientific American. Published monthly by Scientific American, Inc., New York. Each issue contains a variety of articles generally written by the researchers themselves. Level of reading is ordinarily not too difficult. Good source for maintaining some contact with a variety of fields.

Segal, E. M., and R. Lachman. 1972. Complex behavior or higher mental process: Is there a paradigm shift? *American Psychologist* 27: 46–55. A discussion of a variety of causes of paradigm shifts.

Seligman, K. [1948] 1971. *Magic, supernaturalism and religion.* New York: Pantheon Books. A fascinating look at some of the beliefs that have held sway. Unfortunately we haven't gotten over many of them.

Shigeru, N., D. L. Swain, and Y. Eri, eds. 1974. *Science and society in modern Japan.* Cambridge: MIT Press. A wide-ranging series of articles on the development and problems in Japanese science. A number of the observations are quite surprising to those of us brought up in Western tradition.

Silk, J. 1980. *The big bang.* San Francisco: W. H. Freeman. A very readable account of our present knowledge of origin and evolution in our universe.

Singer, P. 1977. *Animal liberation.* New York: Avon. A statement by a philosopher who favors discontinuing the use of nonhumans in research. Long on indignation, short on alternatives.

Skeptical Inquirer. Published by the Committee for the Scientific Investigation of

Claims of the Paranormal, Buffalo, New York. A journal that publishes articles evaluating many of those areas that tend to be considered strange and wondrous. Some are identified as pseudoscience. It has articles on ESP, UFOs, biorhythms, von Daniken, astrology, and many other subjects.

Snow, C. P. [1934] 1958. *The search.* New York: Scribner's. A novel about scientists by an author who is intimately familiar with their world. They're just human.

Snow, C. P. 1963. *The two cultures and a second look.* New York: Cambridge University Press. The original essay examining the scientific and literary points of view, together with some later comments. An enjoyable classic.

Statistics: A guide to the unknown, 2d ed. 1978. San Francisco: Holden Day. Edited by a group of distinguished statisticians, it emphasizes the use of statistics in making scientific decisions—and does it with very little math. Strongly recommended for almost anyone.

Tullock, G. 1966. *The organization of inquiry.* Durham: Duke University Press. Raises an interesting question: How do largely individualistic scientists manage to contribute to an endeavor that is essentially cooperative?

Velikovsky, E. 1950. *Worlds in collision.* Garden City: Doubleday.

Velikovsky, E. 1952. *Ages in chaos.* Garden City: Doubleday. These two books present the author's position that many of the myths of the world, in particular the biblical ones, can be explained by certain violent activity among the planets. The data seem to be judiciously selected to make his case.

von Daniken, E. 1969. *Chariots of the gods? Unsolved mysteries of the past.* New York: Putnam.

von Daniken, E. 1970. *Gods from outer space.* New York: Putnam. Two books that claim that many of the artifacts of the ancient civilizations could not have been produced without the help of mysterious visitors from outer space. The author ignores all offered explanations.

Watson, J. D. 1980. *The double helix.* G. S. Stent, ed. New York: Norton. Includes the text, commentary, reviews, and original papers. Gives a personal perspective of the social and intellectual dynamics leading to one of the most important discoveries of the twentieth century. The other papers put Watson's interpretation into perspective. Highly recommended.

Weart, S. R. 1979. *Scientists in power.* Cambridge: Harvard University Press. The development of nuclear energy. Much of the emphasis is on work by the French. One conclusion is that prior to the Manhattan Project, the United States was behind England, France, Germany, and Russia.

Weisskopf, V. F. 1972. The significance of science. *Science* 176: 138–146. An interesting discussion of the plight of science; includes views both hostile and favorable to science.

White, A. D. [1895] 1955. *A history of the warfare of science and theology.* New York: Braziller. Gives a lengthy and somewhat tedious recounting of the long and at times bloody struggle to separate science and dogmatic theology.

Woolf, H., ed. 1964. *Science as a cultural force.* Baltimore: The Johns Hopkins University Press. Essays on the relations and problems between science and society. Brief but informative and important.

Wyers, E. J., H. E. Adler, K. Carpen, D. Chiszar, J. Demarest, O. J. Flanagan, Jr., E. V. Glaserfeld, S. E. Glickman, W. A. Mason, E. W. Menzel, and E. Tobach. 1980. The sociobiological challenge to psychology on the proposal to 'cannibalize' comparative psychology. *American Psychologist* 35: 955–979. Several psychologists consider Wilson's claim that comparative psychology will be absorbed by sociobiology. Not surprisingly, they don't believe it.

Yearbook of Science and the Future. Britannica Book of the Year *(Encyclopaedia Britannica).* A source of interesting essays on current science.

Index